Clinical Safety and Efficacy Testing of Cosmetics

COSMETIC SCIENCE AND TECHNOLOGY SERIES

Series Editor
ERIC JUNGERMANN
Jungermann Associates, Inc.
Phoenix, Arizona

Volume 1: Cosmetic and Drug Preservation: Principles and Practice, *edited by Jon J. Kabara*

Volume 2: The Cosmetic Industry: Scientific and Regulatory Foundations, *edited by Norman F. Estrin*

Volume 3: Cosmetic Product Testing: A Modern Psychophysical Approach, *by Howard R. Moskowitz*

Volume 4: Cosmetic Analysis: Selective Methods and Techniques, *edited by P. Boré*

Volume 5: Cosmetic Safety: A Primer for Cosmetic Scientists, *edited by James H. Whittam*

Volume 6: Oral Hygiene Products and Practice, *by Morton Pader*

Volume 7: Antiperspirants and Deodorants, *edited by Karl Laden and Carl B. Felger*

Volume 8: Clinical Safety and Efficacy Testing of Cosmetics, *edited by William C. Waggoner*

Other Volumes in Preparation

Sunscreens: Development, Evaluation, and Regulatory Aspects, *edited by Nicholas J. Lowe and Nadim A. Shaath*

Clinical Safety and Efficacy Testing of Cosmetics

edited by

William C. Waggoner

Tewksbury Research
Lebanon, New Jersey

MARCEL DEKKER, INC. New York and Basel

Library of Congress Cataloging-in-Publication

Clinical safety and efficacy testing of cosmetics / edited by William
C. Waggoner.
p. cm. -- (Cosmetic science and technology series ; v. 8)
ISBN 0-8247-7941-X
1. Cosmetics--Toxicology--Testing. 2. Cosmetics--Analysis.
3. Product safety. I. Waggoner, William C. II. Series.
RA1270.C65C55 1990
668'.55'0289--dc20 89-27492
CIP

This book is printed on acid-free paper.

MARCEL DEKKER, INC.
270 Madison Avenue, New York, New York 10016

Current printing (last digit):
10 9 8 7 6 5 4 3 2 1

PRINTED IN THE UNITED STATES OF AMERICA

About the Series

The Cosmetic Science and Technology Series was conceived to permit discussion of a broad spectrum of current knowledge and theories of cosmetic science and technology. The series is made up of books written by a single author and edited volumes with a number of contributors. Well-known authorities from industry, academia, and government have participated in writing these books.

The aim is to cover the many facets of cosmetic science and technology. Topics are drawn from a wide range of disciplines ranging from chemical, physical, analytical and consumer evaluations to safety, efficacy, and regulatory questions. Organic, inorganic, physical, and polymer chemistry, emulsion technology, microbiology, dermatology, and toxicology all play a role in cosmetic science. There is little commonality in the scientific methods, processes, or formulations required for the wide variety of cosmetics and toiletries manufactured. Products range from hair care, oral care, and skin care preparations to lipsticks, nail polishes and extenders, deodorants, body powders, and aerosols to such over-the-counter products as antiperspirants, dandruff treatments, antimicrobial soaps, and acne and suntan products. Thus cosmetics represent a highly diversified field with many subsections of science and art; indeed, even in these days of high technology much art and instinct is used and needed in the formulation, evaluation and selection of cosmetic products, although there is a strong move toward the scientific method particularly in such areas as claim substantiation, product evaluation, and analyses.

Emphasis is placed on reporting the current status of cosmetic technology and science in addition to historical reviews. The series includes books on safety, efficacy testing, oral hygiene products, sunscreens, deodorants and antiperspirants, hair care, and claim substantiation. Contributions range from highly sophisticated and scientific treaties, to primers, practical applications, and pragmatic presentations. Authors are encouraged to present their own concepts as well as established theories. They have been asked not to shy away from fields that are still in a state of development or transition, nor to hesitate to present detailed discussions of their own work. Altogether, we intend to develop in this series a collection of critical surveys by noted experts covering most phases of the cosmetic business.

The eighth book in this series, *Clinical Safety and Efficacy Testing of Cosmetics*, is edited by Dr. William C. Waggoner. The book is divided into two main sections: the first section describes general testing procedures for all the key safety tests used by the cosmetic industry, such as irritation, sensitization, photoallergy, and comedogenicity. In addition to providing the methodologies, it also gives pragmatic advice on how to select outside testing laboratories, criteria of quality for these facilities, and development of strategies for setting up meaningful biophysical tests. The second section discusses the methodologies to determine efficacy of various cosmetic product categories, such as moisturizers, deodorants, antiperspirants, sunscreens, bubble baths, mouthwashes, douches, hair care products, and antiaging products. The techniques discussed range from rather commonplace evaluation methods to the latest sophisticated instrumental methods. These play an important role in comparing different formulations and in claim substantiation programs.

Finally, I want to thank all the authors and editors who are participating in this series, the editorial staff at Marcel Dekker, Inc., and above all, my wife, Eva, without whose editorial help and constant support I would never have undertaken this project.

Eric Jungermann, Ph.D.

Foreword

The physical key to cosmetic usage is product performance. The psychological key is an enhanced sense of well-being.

Life's priceless riches are our five senses, with which most of us are endowed by right of birth. Three of the five, smell, sight, and touch, find expression in cosmetic usage. Our instinctive desire to gratify our senses explains why cosmetic usage dates back to earliest historical records. It also explains its present popularity and forecasts continued growth.

Primitive formulas of natural substances stirred over an open hearth led to household recipes cooked over a kitchen stove. By the eighteenth and nineteenth centuries household recipes formed the basis of a cottage industry. The explosive transition from small business to big business took place in the early decades of the twentieth century. The emergence cannot be attributed to just one factor, but resulted from the convergence of many events, some of which are described below.

Maurice Levy of Park and Tilford Import Company told me that he was responsible for introducing paste rouge into the United States about 1900. In 1914 he commissioned Scoville, a manufacturer of small metal parts, to produce the first lipstick case. Prior to that time rouge was applied to the lips by rubbing a finger over a pat of paste rouge. Gifts of French perfumes brought back after World War I by returning soldiers for their mothers, wives, and sweethearts led to the establishment of U.S. licenses by French perfume companies a year or two later.

In the 1920s and 30s the burgeoning chemical industry began to develop new compounds that could be used in large volume in other industries and, fortunately, some found use in the small cosmetic industry. Four of these have added immeasurably to the cosmetic industry: fluorocarbons, polyvinyl pyrrolidone, synthetic emulsifiers, and plastics.

Fluorocarbons were developed by du Pont as an improved refrigerant and were destined to make air conditioning possible. However, their initial use was as a propellant for dispensing DDT to delouse soldiers during World War II. After the war an early civilian use was as a propellant in aerosol fragrance flacons.

Polyvinyl pyrrolidone (PVP) was developed by the Germans as a blood substitute for treating their wounded soldiers during World War II. After the war the General Aniline and Film Company in the U.S. found a new use for the stockpiled chemical. PVP became the active ingredient in aerosol hair spray with fluorocarbons as the propellant.

Synthetic emulsifiers were another category of compounds developed by the chemical industry. The target market in that instance was the giant detergent companies. However, cosmetic manufacturers were quick to adapt them to improve texture and stability in cream and lotion compounds.

Plastics use was primarily targeted at the container industry, offering unbreakable lighter-than-glass containers which would reduce shipping costs and have greater versatility in shape and design.

The transition from artisan to cosmetic chemist was mandated by the greater scientific knowledge required to formulate products using these synthetic compounds; however, emergence of the cosmetic chemist is surprisingly recent. The first Society of Cosmetic Chemists worldwide was founded in New York City in 1945 with 12 members, and membership rose to 3300 by 1985.

This growth in the cosmetic industry and its personnel was also due largely to the discovery of new molecules that enhanced product efficacy. As some new molecules were introduced, it was noted that the formulation did not hold together and/or the products caused irritation or sensitization reactions. This led to the need for predictive testing for efficacy and safety. We began testing products for their acceptability and for their propensity to cause reactions. In some of the first tests we placed the product on cotton balls and taped them to the skin. This method eventually evolved into the modern day "patch" using occlusive bandages. We also found that when some of the standard natural compounds underwent minor chemical modification, they retained their cosmetic properties but decreased their irritation and sensitization potential.

Testing needs grew when distribution mushroomed from thousands to millions. As the number of users increased, so did the possibility of consumers that might find the product unacceptable from an efficacy or safety perspective. Company executives wanted more predictive forecasts of field problems to circumvent consumer dissatisfaction and potential liability. This particular volume is timely in that it places in a single text much of the accumulated knowledge on clinical testing of cosmetics. It will be a cornerstone reference for many of those in cosmetics research and development.

Let me share a personal anecdote relating to most cosmetic marketers' attitude toward research and development in the early 1940s. When the president of a nationally known cosmetic company of that period asked me why I had sought an interview with his company, I replied that I wished to be employed there as a research chemist. He looked down on me and commented that after the war all the women would go back to their kitchens and the prospering cosmetic industry would collapse. Several years later when I was president of my own company this same person applied to me for a job. As we met he said that I looked vaguely familiar. I never told him of our initial meeting.

Hazel Bishop
Fashion Institute of Technology
New York, New York

Preface

The initial purpose of this book was to serve as a wide-spectrum resource volume for cosmetic clinical research. With the cooperation of many contributing authors, the book began to take shape. Every field of cosmetics is represented here, including some products that are considered to be drugs by virtue of their being a vehicle for various therapeutic agents. Because of the proprietary nature of the exact testing methods and operational procedures of many companies, some authors were reluctant to disclose all of the details involved in their work. The guidelines and protocols presented here are nevertheless sufficient to help the reader adapt and formulate an appropriate procedure to fit whatever type of product is to be tested.

The contributors have worked diligently to produce this compendium. They are all excellent scientists who responded well to goading, pushing, shoving (and sometimes encouragement) to get their sections completed. The reader will reap the rewards.

Dr. Maurits Dekker was an initial inspiration to this volume during one of our several conversations on Dutch literature. Sandra Beberman used her infinite patience in her encouragement to help us all continue the task. Hazel Bishop applauded the project and offered her own insights of the early days of clinical testing of cosmetics. Finally, I give the authors a multitude of praise for listening to my guidance (however inept at times) to complete their chapters and to share their knowledge with their reader colleagues. Many thanks to all of you.

William C. Waggoner, Ph.D., FAACT

Contents

Contributors

RICHARD S. BERGER University of Medicine and Dentistry of New Jersey, Robert Wood Johnson Medical School, New Brunswick, and Hill Top Research, Inc., East Brunswick, New Jersey

WARD L. BILLHIMER Clinical Safety, Hill Top Research, Inc., Cincinnati, Ohio

JAMES P. BOWMAN Hill Top Research, Inc., East Brunswick, New Jersey

LEWIS P. CANCRO* Personal Products Division, Lever Brothers Company, Edgewater, New Jersey

KIM S. DAMMERS Hill Top Research, Inc., East Brunswick, New Jersey

ALAN S. DANA, JR. Clinical Research Unit, Techni-Med Consultants, Inc., Wyncote, Pennsylvania

*Current affiliation: Personal Products Research and Development, Chesebrough-Pond's, Inc., Trumbull, Connecticut

HARRY R. ELDEN Institute for Skin Research, Xienta, Inc., Bernville, Pennsylvania

ANNA K. FLOYD Product Safety and Claims Substantiation, Mary Kay Cosmetics, Inc., Dallas, Texas

GARY L. GROVE KGL's Skin Study Center, Broomall, Pennsylvania

S. MARK HENRY Bristol-Myers Products, Hillside, New Jersey

EDWARD M. JACKSON Noxell Corporation, Hunt Valley, Maryland

MARVIN KAMINSKY Toxicology Group, Beecham Products, Parsippany, New Jersey

ALAN D. KIRSCH Medical and Regulatory Affairs, Beecham Products, Parsippany, New Jersey

LINDA A. LAZAR Consultant, Bloomfield, New Jersey

DAVID L. MILLER Product Safety and Claims Substantiation, Mary Kay Cosmetics, Inc., Dallas, Texas

OTTO H. MILLS, JR. University of Medicine and Dentistry of New Jersey, Robert Wood Johnson Medical School, New Brunswick, and Hill Top Research, Inc., East Brunswick, New Jersey

CHARLES J. MINNICH Techni-Med Consultants, Inc., Wyncote, Pennsylvania

BERNARD MISEK Medical and Regulatory Affairs, Beecham Products, Parsippany, New Jersey

ANTHONY J. PARISSE Product Evaluation, Carter-Wallace, Inc., Cranbury, New Jersey

PETER T. PUGLIESE Peter T. Pugliese, M.D. and Associates, Bernville, Pennsylvania

THOMAS J. STEPHENS Product Safety and Claims Substantiation, Mary Kay Cosmetics, Inc., Dallas, Texas

WILLIAM C. WAGGONER Tewksbury Research, Lebanon, New Jersey

Introduction

In 1959, I first became involved with the safety and/or effectiveness testing of cosmetic, toiletry, and other products that fall under the Federal Food, Drug, and Cosmetic Act definition as "Articles intended to be rubbed, poured, sprinkled, or sprayed on, introduced into, or otherwise applied to the human body or any part thereof for cleansing, beautifying, promoting attractiveness or altering appearance." I spent a considerable amount of time in the library trying to find information on how to conduct adequate testing for the safety and effectiveness of a new cosmetic product. The only information on safety testing I came across was a compilation of papers on "Appraisal of the Safety of Chemicals in Foods, Drugs and Cosmetics," prepared by FDA toxicologists. Efficacy testing was only a glimmer in people's minds. Waving lotion waved hair, hair dyes dyed hair, moisturizing cream moisturized, lipstick colored lips, and rouge colored cheeks. The products were simpler, and if consumers liked a product they continued to purchase it.

The technical and legal literature of the 1920s and '30s, however, makes it apparent that questions had been raised about the safety of cosmetic products and the claims being made for some products. There had been Congressional inquiries into why some cosmetic products had produced various adverse reactions on humans and why these adverse reactions could not have been ascertained prior to marketing these products to the public. Because of these inquiries cosmetics became regulated in 1938, when the Federal Food, Drug, and Cosmetic Act was enacted to define "cosmetics" and thus bring them under regulatory control. This regulatory control was tightened with the enactment of the Color Additive Amendments in 1960 and with FDA promulgation of its regulation requiring adequate substantiation of safety in 1975. Substantiating the safety of cosmetics is described in Chapter 26 of *The Cosmetic Industry: Scientific and Regulatory Foundations* (Dekker, 1984), edited by Norman Estrin.

Although cosmetic safety testing has since the 1950s been in the forefront, testing for effectiveness lurked in the background. Effectiveness studies have been conducted for marketing purposes, to ensure that the consumer perceives the claimed effects. Unlike the situation for drugs, proof of effectiveness is not required for cosmetics. Nonetheless, it must always be kept in mind that the claims made for a cosmetic product can cause one to cross that thin and ill-defined line between a cosmetic and a drug, particularly as science marches forward. As new scientific knowledge is obtained about the action mechanism of truly effective cosmetics, cosmetic and drug product categories come ever closer together.

Since 1959, the medical sciences, and especially the disciplines of dermatology and toxicology, have developed well-established testing techniques. Standardized methods for evaluating, for example, the potential irritancy of an ingredient and/or product are fairly well delineated for the skin and eye. In many instances, these testing protocols have been incorporated into both domestic and international health regulations and directives. Many studies evaluating the accuracy of irritancy studies have been published in the technical literature. Discussion of the positive and negative aspects of the testing protocols has been conducted at technical meetings and seminars.

Numerous studies have been conducted and published on evaluating human irritancy, sensitization, and photosensitization of an ingredient and/or product. Certainly, our scientific knowledge has been enhanced and we have improved our ability to evaluate the safety and effectiveness of products. This is important in light of changes in our industry and our products.

During the 1970s, the Food and Drug Administration conducted an intense review of the many active ingredients used in over-the-counter (OTC) drug products. These ingredients were broken down into specific categories of product use. The safety and effectiveness of each ingredient were evaluated by panels of experts. These panels laid down the basic foundations for what an expert third party would regard as evidence that substantiated the safety and effectiveness of an OTC cosmetic drug. Either these foundations for protocols were accepted by the panel from data submitted by the industry, or the protocols were enhanced by the panels so as to ensure, in their judgment, adequate data to substantiate effectiveness.

One cosmetic industry, through its trade association, the Cosmetic, Toiletry, and Fragrance Association (CTFA), has spent a considerable amount of time and money in gathering the past and present technical literature to establish the safety of ingredients used in cosmetic products. This literature has been reviewed by an expert panel, the Cosmetic Ingredient Review (CIR) Panel, which has either accepted or rejected the ingredients as safe for the use intended based on the data submitted, or has required that additional studies be conducted. The CIR panel has performed this function using procedures modeled on the FDA OTC drug review. Results of the CIR panel's findings may be found in the *Journal of the American College of Toxicology*.

Thus, today, established protocols exist for evaluating cosmetic products to ensure adequate substantiation of both the safety and the effectiveness of a product prior to marketing, unlike 1959. Many of these protocols and their rationale are discussed within the covers of this book.

Robert P. Giovacchini, Ph.D.
The Gillette Company
Rockville, Maryland

Part I

GENERAL TESTING

1

Independent Clinical Laboratories: Qualification, Study Placement, Study Monitoring, and Future Service Trends

Thomas J. Stephens, Anna K. Floyd, and David L. Miller

Mary Kay Cosmetics, Inc., Dallas, Texas

I. INTRODUCTION

Today's cosmetics consumer is a smart shopper. While attractive packaging is still an important factor in product appeal, the cosmetic user expects skin care formulations that are elegant, safe, and live up to the consumer's expectation of effectiveness.

Traditionally, consumer product companies conduct safety studies to ensure products do not produce adverse reactions from normal usage as well as reasonable foreseen misusage, such as getting a shampoo into the eyes or a child drinking perfume. Similarly, claims substantiation departments conduct various in vitro and clinical studies to substantiate the functionality of the product. Current trends have shown that product claims often determine brand selections. For this reason, the results of clinical studies with cosmetics have become important marketing tools. A trip to the department store shows that science sells cosmetic products. With many products, efficacy claims and supporting technical data are displayed in prominent areas of the package to catch the buyer's eye. The reader of this information not only includes the discriminating consumer, but also competitors, consumer groups, and the U.S. Food and Drug Administration (FDA).

Product claims can be categorized into three general types: pre-emptive, comparative, and more recently cosmeceutical. Pre-emptive claims emphasize product aesthetics, functionality, or safety without comparing the product to some standard or competitive product. For example, the statement that a product is "clinically tested for irritation, allergy, and comedogenicity by an independent laboratory under a leading dermatologist" communicates a fact that the product was evaluated by a physician specializing in the skin, without stating the outcome or that the product is appropriate for every individual in every situation. A consumer reading this claim would probably expect this product to be safe, and may select this product over a comparable product without a safety claim.

Comparative claims state that a product is at parity or superior to an established bench mark. Clinical results comparing two over-the-counter drug (OTC) therapies such as salicyclic acid and benzoyl peroxide might lead the acne-prone buyer to change brands in order to get the strongest acne-fighting medicine possible. This type of claim often results in public rebuttals and law suits from competitors wanting to retain loyalty to their brands.

Cosmeceutical claims are a type of pre-emptive claim which associate product usage with some desirable physiological skin change such as the elimination of wrinkles. Some cosmeceutical product claims are not adequately substantiated using scientific methodology, but depend upon testimonials from celebrities as evidence of efficacy. Recently, the FDA has taken the stand that many cosmeceuticals should be considered drugs because of their treatmentlike claims [1].

Industrial scientists faced with the newly acquired task of assessing the safety and efficacy of products may have limited experience in selecting investigators and clinical sites for studies. Seasoned toxicologists and claims substantiation scientists often provide the names of qualified laboratories and medical centers to junior members without giving the rationale for the selection process. This chapter has been written to help develop competency in these areas. The contents of this chapter discuss qualification criteria for independent contract laboratories, study placement, and study monitoring, and concludes it with a brief discussion of new service trends which are likely to develop during the next decade.

II. QUALIFICATION OF A CLINICAL LABORATORY

It is becoming increasingly apparent, in the competitive field of cosmetics, that efficacy and safety claims are powerful product marketing tools. As a wider variety of innovative glamour and skin care products are developed, the responsibility for substantiating the associated claims is often overwhelming for the average company. As a result, many cosmetic manufacturers are relying heavily on contract laboratories to conduct the studies which will help predict the success and integrity of each new product. Conducting studies out-of-house, however, does not diminish the manufacturer's responsibility for obtaining quality data based on scientific fact. For this reason, the selection of a reliable contract laboratory is fundamental in the testing process. A consistent, thorough, systematic approach to qualifying a new contract testing facility helps to significantly reduce any risk involved in placing a study with an independent group. A simple checklist (Table 1.1) has been developed as a guide through the qualification process. The points highlighted in Table 1.1 are discussed in greater detail in the subsequent text. The qualification process described is similar to a general procedure published by Dr. E. M. Jackson [2]. This basic approach has been expanded to provide a step-by-step system for qualifying a contract laboratory.

A. Prescreening Procedure

To expedite the process of initially identifying contract laboratories, one should first explore several available resources. For example, most contract laboratories distribute advertising brochures at scientific meetings where their specific areas of expertise are relevant. Often a representative will be present to personally expand upon the printed information provided. A related source of information can be found in scientific journals. Articles describing clinical studies will generally cite the participating investigators or laboratories. This source of information often provides more detailed, first-hand examples of the quality of science represented by a testing group. Supplemental guidance by word of mouth from acquaintances in your industry is a third effective resource. More specific, candid opinions than those that would typically be found in published material can usually be elicited from peers. As the above information is acquired, it should be organized, dated, and retained in a permanent file.

TABLE 1.1 Qualifying a Contract Laboratory Activity Checklist

I. Preparation of general outline of study objectives prior to contacting potential contract laboratories

Determine specific needs in terms of facilities, instrumentation, and clinical expertise.

Develop profile of desired panelist (patient) population requirements.

Estimate budgetary limitations for study.

II. Prescreening of several laboratories by phone or letter

If available, review material available on file describing testing facilities.

Select potential sites based on specialties, staff size, equipment, and patient populations that are compatible with proposed study.

Contact lab; discuss general parameters of study and request additional information.

Name of principal investigator(s)
Hospital or academic affiliations
Names of other clients
Sample protocols

III. Initial site visit

Inspect facilities and equipment; check calibration records.

Review standard operating procedures for compliance with good GCPs and GLPs.

Examine records management system.

Facility should have adequate space to maintain study records completely and confidentially for each client for 2 years.

Observe data entry process.

To minimize bias, an objective recorder, other than the investigator, should make entries on CRFs. If available, computerized records should be compatible with sponsor's system.

Meet with coinvestigator and staff coordinator to discuss protocol.

Check availability of suitable patient population, number of panelists needed, and demographic requirements.

Discuss preparation of CRFs and informed consent. Inquire about accessibility of IRB.

Observe general organization, confidentiality, and professionalism maintained during visit.

Prepare a formal report summarizing impressions to place on permanent file.

IV. Trial run prior to official study.

Send a sample (previously tested by a reliable laboratory) for a brief evaluation.

Compare execution, timeliness, and accuracy to earlier results.

Make a decision on the basis of the outcome of steps II, III, and IV.

V. Follow-up communication after placement of study.

If possible, visit the laboratory at start of study to monitor the following activities:

Adequate patient recruitment
Proper completion of CRFs, accurate baseline data collection, correct randomized assignment, and distribution of product
Complete instructions for panelist compliance

If necessary (particularly for long-term studies), make a second visit to monitor progress and review CRFs.

Prepare formal report of results of each trip to the site.

After compiling a list of potential investigators, several brief phone conversations with the respective laboratory directors will usually serve to further narrow the field. During the course of the interview, particular emphasis should be placed on the reputation, credentials, and experience of the staff investigators. It is essential to match technical expertise with the product testing demands in order to ensure that studies will be competently run. As a general rule, cosmetic manufacturers should select contract laboratories which emphasize dermatological and ophthalmological specialties, as opposed to internal medicine, which would be a more desirable specialty for a systemic drug study. An additional advantage of selecting an experienced practicing dermatologist or ophthalmologist is his or her accessibility to various panelist populations. It will generally simplify the process of locating patients with a specific skin condition if the participating dermatologist is able to identify subjects from his or her own private practice.

Other factors influencing the decision process during this prescreening phase include geographic location, scheduling, and supporting staff size. Restrictions on these criteria should be outlined in advance and discussed with the laboratory personnel. When a prospective match is found the sponsor should request that the contract laboratory provide documents with the additional detailed information outlined in the checklist in Table 1.1.

B. Initial Site Visit

Following a review of the requested material, a site visit should be arranged to evaluate the laboratory. The following general objectives are recommended for an initial visit:

Introduction of administrators, coordinators, and investigators
Inspection of facilities, equipment, and calibration records
Observation of standard operating procedures (SOPs) for compliance with good clinical practice (GCPs) and good laboratory practice (GLPs) regulations
Examination of records management system
Assessment of the professionalism and overall interaction of the laboratory staff

A tour of the facility is the quickest and most effective means of assessing the suitability of a laboratory for the specific requirements of the cosmetic manufacturer. To ensure adequate access to key personnel, resources, and all pertinent information, arrangements should be made and confirmed in advance of the visit. Upon arrival at the laboratory, a recommended starting point is to inquire about administrative matters such as SOPs, regulatory letters on file, records management, and product accounting procedures.

It is important that complete SOPs be on file for all procedures advertised by the contract laboratory. These documents should be signed and dated by appropriate personnel and officers. For additional evidence of compliance with GLPs and GCPs, it is appropriate to inquire whether the laboratory has received any regulatory letters. The presence of these forms on file is documentation of problems or observations cited during an FDA investigation and should be viewed as a possible indication of weakness in compliance with GCPs or GLPs. If the problems have been resolved, these actions should be documented and available for review.

In regard to records management, adequate space should be provided for safe, long-term retention (defined by GLPs as 2 years after completion of the study) of each client's file. More specifically, these documents should be maintained in a separate area which is secure, with access limited to authorized personnel. This confidentiality is essential for the protection of the manufacturer who must maintain a marketing edge by not giving the competition foreknowledge of new product introductions. At no time should a client's records be made available to a visitor or monitor from another company. Any violation of this confidence observed during a site visit represents a serious lack of professionalism on the part of the contract laboratory. Another area requiring an extra degree of security involves the storage and accountability procedures for test products. Inadequate accountability for clinical inventory is probably indicative of a slack approach to product assignment and distribution, which could directly impact the results of a study.

In addition to the hard copy files discussed above, a computerized data entry system is a useful tool offered by many testing facilities. If the system is compatible with that of the sponsor company, data can be loaded and transferred long distances immediately. Information handled in this manner can be easily analyzed statistically and charted or plotted directly on the computer. There are certain standards to watch for, however, when evaluating the merit of an automated records management system. The contract laboratory must maintain tight quality control, validating data input and verifying existing data on a regular basis. As with all equipment and procedures, there should be SOPs for maintenance of the computer, with updated signed documentation verifying compliance with GLPs. Some systems are equipped with built-in safeguards which signal the investigator when deviations are made from the protocol. Although this is a nice feature, a more necessary safeguard is one that prohibits the investigator from editing the data once it is transferred. As a final point regarding automated data entry, the visiting monitor should inquire about the laboratory's policy on maintaining hard copy case report forms (CRFs) as back-ups for the entered data. This is important for quality control when raw data is recorded by hand on the CRFs, then entered onto the computer at a later time.

An initial visit is also the time to discuss administrative matters such as whether the contract laboratory provides their own informed consent forms, CRFs, protocols for routine procedures and Institutional Review Board (IRB) approval where needed. If these forms are available, request sample copies for review. Case report forms are usually individualized to accommodate the data of a particular study and therefore will not be discussed at this time. Informed consent forms, on the other hand, are closely regulated and are legally required to contain the following components [3]:

A statement that the study involves research, with an accompanying explanation of the purpose of the research.
A description of the procedures involved and identification of which procedures are experimental.
An estimation of the duration of the subject's participation.
A description of any benefits to the subject which may result from the study.
Information concerning any foreseeable risks to the subject.
If more than minimal risk is involved, a statement regarding what compensation or medical assistance will be made available to the panelist. (Patients should be provided with a name and number to call if an adverse reaction should occur.)
A statement describing to what extent confidentiality of records will be maintained.

In addition to the above, the consent form should assure the subject that participation is voluntary and that withdrawal from the study will not result in loss of any benefits to which the subject is otherwise entitled. The written consent form should be approved by an IRB, which is independent of the investigator. The FDA has provided guidelines for the responsibility, structure, operation, and accountability of a duly constituted IRB. In terms of organization, the board must include a minimum of five members, representing both sexes and more than one profession. At least one member must be from a nonscientific field. Generally, the function of the IRB is to evaluate research protocols on the basis of subject risk, adequate informed consent, proper protection of the privacy of the subject, appropriate documentation, and adequate monitoring of the data collected. The IRB is required to retain documented records of its interaction with an investigator for a period of 3 years.

During the physical tour of the facility, keep track of available instrumentation and question applicability for specific measurement needs. Discuss equipment capabilities with technicians if possible, and request calibration data on instrumentation. These data should be readily available in log books. Entries should include current documentation of general maintenance, repair, and calibration, with dates and signatures.

In addition to the above, an initial site visit provides an excellent opportunity to evaluate the overall atmosphere of the contract laboratory. An attitude of cooperation, efficiency, and unity among members of the staff often reflects a prevailing spirit of professionalism and organization which is a significant indicator of the quality of studies conducted by the laboratory. Some positive attributes to look for include the routine assignment of project leaders to coordinate all aspects of a particular study, a nucleus of full-time employees who will be available for communication with the sponsor throughout the study, one or more recruiters familiar with histories of compliance of panelists, and technical personnel aware of the importance of accuracy and adherence to protocol.

A formal report of the trip should be included in an individual file developed for the specific contract laboratory. For adequate documentation of a site visit for the purpose of qualifying a lab or monitoring a study, always record the following:

Laboratory name and address
Monitor's name
Date of visit
Objectives of trip
Summary of findings
Statement of action taken or modification to protocol (for study in progress)
Attachment of completed qualification checklist

Any additional information or phone conversations should be documented and retained in this file whether or not the laboratory is selected to conduct the study.

III. FACTORS INFLUENCING STUDY PLACEMENT

A. The Use of Scientific Expertise In-House Versus Out-of-House

Table 1.2 lists seven major factors influencing the placement of clinical studies. A review of the scientific expertise available in-house and out-of-house should head the top of the list for individuals placing a clinical study. Defining the responsibilities of a technical staff with diverse talents is crucial for an efficient operation. In most companies, product safety and claims substantiation management is expected to utilize scientific expertise in the best possible way so as to meet the mission of the company. In turn, the role of the safety and claims

TABLE 1.2 Factors Influencing Study Placement

Scientific expertise—in-house versus contract laboratory
Compliance to good laboratory and clinical practice regulations
Cost
Geographic location and availability of subjects
Specialized equipment requirements
Timing of study placement
Proprietary formulations
Special considerations in claims substantiation

substantiation staff is to select and place tests, ascertain the integrity of the work through diligent monitoring of the study, and extrapolate study data into a safety and claims assessment on products.

If the corporate environment is one of cost savings and increased employee productivity, it is generally more effective for companies to place clinical studies out-of-house. The technical staff can focus their time on designing good safety and claims product programs, exploring new procedures, and supporting the research and development process.

In deciding where to place a clinical study out-of-house, consideration should be given to the credentials, reputation, publication record, and experience of the investigators [4]. In the actual placement, emphasis should focus on matching the technical competencies of the investigator with the special requirements of the clinical study. For example, using the combination of an ophthalmologist and dermatologist as coinvestigators on a clinical study of an eye area product would result in a more thorough investigation than using either medical specialist alone.

It is the scientific expertise of the investigator that lends credibility to the protocol design, clinical observations on the case report forms, and final study report. As a word of caution, some laboratories have business managers designing nonroutine clinical studies. While no one would doubt the years of experience they may have had in the field, their role should be limited to their areas of expertise such as scheduling studies and quoting prices. The protocol should be designed and the study supervised by an experienced individual with either a doctorate in medicine (MD), dental surgery (DDS) or philosophy (PhD), with the latter being earned in a biomedical discipline.

When making a selection, remember the investigator for your clinical today could be your expert witness in a liability case tomorrow.

In a limited number of situations, it is not desirable to place a study out-of-house. These situations include the use of unique technology or scientific talent unavailable at contract laboratories, time constraints imposed by product development deadlines, and the rapid screening of multiple prototype formulations for safety and efficacy prior to their first-time use by a limited number of panelists.

B. Compliance to Good Laboratory and Clinical Practice Regulations

Whether testing cosmetics or OTC drugs, the laboratory you select should comply with the requirements of good laboratory practice and clinical practice regulations. These laws were passed to ensure that adequate record keeping and organizational systems are applied to preclinical and clinical studies [5]. Safety studies should only be placed at laboratories complying with these regulations.

C. Cost

For toxicologists and claims scientists operating with a tight testing budget and a large number of products to be evaluated, the cost of safety and efficacy testing becomes important. In the long run, it is always better to select qualified laboratories with experienced investigators rather than unknown laboratories with lower prices. If the price seems high at a reputable laboratory, consider simplifying the study's design and/or reducing the number of test groups without compromising the protocol's objective. Most laboratories negotiate the price of large studies and give price discounts to clients pledging a certain dollar business per year.

On a dollar per dollar basis, the cost of conducting clinical studies in-house is less expensive and can save up to 60-70% on the price of most procedures. However, if this course of action is followed, the department's technical staff will spend all their time recruiting subjects and serving as principal investigators on studies, and may have a difficult time meeting other departmental and corporate goals. As previously mentioned, studies conducted at independent clinical laboratories under the direction of reputable investigators lend a credibility to the findings which is difficult to duplicate by conducting the study in-house.

A middle ground to this situation can provide an optimized use of in-house and contract lab capabilities. This involves direct participation of the sponsor's scientists in critical measurement procedures in an out-of-house study. High-tech equipment is provided and operated by the sponsor at the contract lab site. The outside contractor provides the overall supervision and execution of the study, including the panelist recruitment, questionnaire administration, product distribution, and general record keeping.

D. Geographic Location and Availability of Subjects

Geographic location is an important factor when testing some OTC drugs and moisturizing cosmetics. Cold dry weather exacerbates acne and xerosis (dry skin). Consequently, winter provides a good environment for conducting efficacy trials with drugs such as benzoyl peroxide, salicyclic acid, and cosmetic moisturizers. In contrast, the hot moist environment of the beach during the summer months is the traditional environment for conducting usage studies with sunscreens. Occasionally to the dismay of the claims substantiation scientist, a product will need efficacy evaluation at the wrong time of the year in order to support a launch date. Assuming it is impossible to wait for the ideal environment, the best course of action is to contact an experienced investigator and to consider various strategies such as conducting the test in a foreign country with the appropriate climate, or taking a chance and conducting the test in a less desirable climate with the appropriate positive and negative controls.

The inability of a clinical laboratory to recruit the desired population can affect the study completion deadline and may indirectly slow down the development of a product. Claims scientists and toxicologists should have a good understanding of a laboratory's ability to recruit a target population of subjects before placing a study.

The need for panelists with special demographic requirements such as age or race may lead to the selection of a laboratory based on its ability to recruit the desired subjects. An example is the ease with which clinical laboratories located in popular sunbelt cities can recruit retired citizens.

E. Specialized Equipment Requirements

Selecting a laboratory with the appropriate high-tech equipment requires some training in the bioengineering field in order to understand the strengths and weaknesses of various pieces of equipment. For example, the application of thermography to demonstrate blood flow to an area will give qualitative but not quantitative information. While thermography makes a powerful marketing tool to visualize a change, it is not as useful and certainly more costly than the application of laser doppler technology to the same problem.

Clinicals requiring the use of specialized equipment which is expensive and or likely to be used infrequently are good candidates for out-of-house placement. Table 1.3 is a partial listing of equipment found at clinical laboratories. The selection of the equipment should be based on the type of biological change(s) anticipated in the study. The resultant measurements should supplement the clinical observations of a trained investigator. Care should be taken to ensure the laboratory staff has adequate knowledge and, in some cases, certification of training with the equipment. Attention should be paid to equipment care by looking at log books for indications of repair and the frequency of calibration. For in-house screening purposes, it might be advantageous to purchase selected equipment such as an evaporimeter, skin impedance, and photographic equipment for rapid evaluation of prototype formulations.

F. Timing of Study Placement

Meeting a product launch deadline may be the overriding factor in choosing a laboratory. Many product development activities feed off the outcome of preclinical and clinical studies. Preclinical study results often serve as product development bench marks that allow the toxicologist to approve in-house panel testing of a prototype product and provide the basis for warning and cautionary statements dealing with accidental ingestion and skin and eye irritation. In some instances, subchronic and chronic tests are conducted to evaluate the potential of a household product to produce serious effects, such as systemic toxicity through skin absorption, carcinogenicity, reproductive and teratogenic anomalies, and toxic responses to the central nervous system. Adequate preclinical studies provide the basis for proceeding with clinical investigations.

A clinical safety program for a topically applied product may include acute and cumulative irritation studies, human allergic

TABLE 1.3 High Tech Equipment

Equipment	Purpose	Cost[a]
Perfusion monitors	Detect microcirculation changes	10
Evaporimeter	Transepidermal water loss, wound healing, barrier effects	8
Elastometer	Mechanical test, skin stiffness	6
Skin impedance	Moisturization	6
Thermography	Detect circulation changes	20
Photographic equipment	Visual record	3
Image analysis	Measure various features: photo, real time, replica	10-30
Surfometer	Measure contour, roughness parameters of surface	6-10
Solar simulator	SPF testing	16

[a]In thousands.

and photoallergic contact dermatitis testing, comedogenicity evaluation, and prolonged usage studies conducted under the direction of a clinical investigator. A clean clinical program triggers the approval of many costly expenditures and starts several labor-intensive activities for a product. These may include purchasing raw materials, product cartons and containers, developing package artwork, and the manufacturing of the product. Starting a test on time is important because of the dependency of later activities in the product development network. In most tight timing situations, the sponsor should select a laboratory with a proven track record. Laboratories that are responsive to a client's needs will have repeat business.

G. Proprietary Formulations

Clinical studies on proprietary formulations without patent protection are good candidates for in-house evaluation. While most laboratories have adequate security safeguards to protect proprietary formulations, the likelihood of a breach in security is greater out-of-house. If in-house capabilities are unavailable, the contract laboratory should sign a confidential disclosure agreement prior to conducting the study. Testing of proprietary formulations should be discussed with senior management and the legal department.

H. Special Considerations in Claim Substantiation

In qualifying a contractor for cosmetic claims substantiation work, several points must be considered in addition to those considered for qualifying a safety test contractor. Claims work frequently involves instrumental and/or expert subjective evaluations in which nonstandard techniques abound. It is therefore important to determine (1) the precise understanding and value of the technique, (2) standardization and/or calibration methods, and (3) the interpretation the contractor places on his measurement. This information can then be compared with literature references provided by the contractor, or preferably searched independently by the client's staff.

The nature of the skin in regard to efficacy measurement often requires controlled conditions under which observations are made. The impact of local environmental factors (temperature, humidity, noise, stress) on the test panelist prior to and during the measurement is often significant when it comes to objectively determining the quality of the skin. If special equipment is to be supplied by the contract lab, the availability of the equipment for pilot evaluation prior to placing the study should be ascertained. Finally, if the technology is being supplied by a third party, their qualifications should be investigated. For example, while the primary test facility may carry out evaluations in an air-conditioned room with facilities for accommodating several panelists without crowding, this may not be the case at the subcontractor's facility.

IV. MONITORING A STUDY IN PROGRESS

An extension of the qualification of a contract laboratory includes the careful monitoring of a study to assure adequate recruitment of panelists, proper completion of CRFs, a close adherence to the protocol, and panelist compliance. Additional visits to the laboratory both at start up and sometime toward the middle of the study are strongly recommended for this purpose, particularly whan a newly qualified laboratory is involved. Attention to detail in documentation of all these areas is critical to the overall acceptability and validity of the study.

The major objective of a visit at start up of a study is to make certain it is properly filled (panelists meet inclusion/exclusion criteria), and to check completion of history and consent forms. The informed consents should be signed and witnessed immediately, with matching dates recorded beside both signatures. The information recorded at baseline is generally reviewed more thoroughly by the FDA for inconsistencies or discrepancies in data than the actual results reflecting the efficacy of the product.

The second, in progress, monitoring trip is advisable for the purpose of reviewing study files, inventorying clinical supplies, resolving CRF discrepancies, and meeting with the investigator. Any problems encountered by the contract lab staff or errors observed by the monitor are more easily rectified at this point than after the final report is issued. This early examination of data may also provide direction for ongoing formulation modifications if adverse trends are noted. If any panelist withdraws from the study because of an adverse reaction, the investigator should record his or her observations along with the symptons described by the patient on an attrition form (provided in the protocol). The monitor should review these forms during the visit and request that the investigator's staff contact the dropped panelist if additional information is required. This type of investigative research is often necessary to pin down the actual source of the reported reaction and to determine if it is indeed product related. Timely follow-up is critical to ensure that contact is made with the panelist before his or her memory of the specific circumstance of the reaction is lost. In addition to these advantages of closely monitored studies, the extra interest expressed by the sponsor lays the groundwork for open lines of communication between the client and contract lab in all subsequent collaborative efforts.

V. POTENTIAL GROWTH FOR CLINICAL LABORATORIES

During the 1970s, the promulgation of federal regulations such as the Toxic Substances Control Act and good laboratory practice and good clinical practice regulations, and the need for additional testing space were important factors in the growth of the contract laboratory business. With today's abundance of contract laboratories, growth depends on anticipating the special needs of the cosmetics, pharmaceutical, and consumer products industries. It is highly unlikely that increased business will come from the expansion of traditional testing methodology such as the human repeated insult patch test, but instead growth will evolve from new services and improved ways of evaluating the safety and efficacy of products.

Long-term growth will parallel the development of new types of raw materials and consumer products. For example, future claims such as specific sun protection from the harmful effects ultraviolet rays will certainly result in the development of new skin protection factor (SPF) testing methodology in order to substantiate this efficacy claim. Future growth will be influenced by the consumers' expanding definitions of produce safety and efficacy. The consumer of tomorrow will expect future skin care products to produce significant perceivable improvements in skin with fewer side effects such as stinging, burning, and itching. Other changes in the business will come as evolutionary thrusts fueled by technological advances in computing, electronic communication, and measurement systems for clinical efficacy and safety testing.

A. Impact of Computers

Microcomputers are dropping in price to levels that no longer preclude even the smallest contract facility from taking advantage of the functions computers can do much better than people: the efficient management of large bases of data and record keeping. Tedium and errors in report generation can be effectively eliminated by merging data base output (computer-generated tables of data) with the report text in a word processing program. As in all computer applications, the quality of the result is ultimately dependent on the quality of the input. In spite of the use of computers to perform the above-mentioned functions, the heart and bottleneck of the evaluation reporting system is still the traditional CRF from which observations must be transcribed and manually entered into the computer. In order to achieve rapid error-free turnaround of study results, the inexpensive and portable lap top microcomputer will eventually find application at this state of the clinical evaluation. These small units bring together the convenience of a data sheet and the reliability of audited data stored in an electronic format, ready for immediate processing. The use of bar codes and bar code readers will also be found an attractive alternative to tracking test samples as well as recording the results of clinical observations by means of encoded questionnaires.

B. Communications

The second thrust in future growth represents perhaps a more radical shift in a way of doing business and involves the sharing of computer resident data generated by the contract laboratory with the sponsor practically as it becomes available. This would expedite the use of contract lab involvement in the screening of formulations during the development stages where quick turnaround is required to maintain an effective application of the testing. Additional labor savings and error reduction is achieved because rekeying of data is not required for statistical study of data by the sponsor.

C. Measurement Instrumentation

The third thrust in future growth we wish to cover here is in the area of application of instrumentation to enhance and expand the objectivity of the clinical measurement. The basis of practically all clinical data in the safety and efficacy testing of cosmetic products has been the trained human observer. This essentially requires the observer to make a qualitative or quantitative decision at the time of the observation. Sophisticated evaluation parameters, such as skin texture and coloring, are more confidently judged with the aid of photographic records which allow for standardization and detailed study. Some parameters such as microcirculation and stratum corneum barrier quality cannot be evaluated at all without the use of instruments. A listing of some of the high-tech instrumental techniques with approximate equipment costs is found in Table 1.3.

Some of this new equipment is useful in a number of different applications (e.g., portable skin reflectance color meters and the Servomed evaporimeter), and therefore is justifiable as part of a clinical lab's equipment roster. Other types of instrumentation are very complex and expensive and/or are more specific in application and likely to be dependent on what's hot in the industry and on the marketing strategies of the sponsor company. In the latter case, the investment by a given contract lab in equipment and expertise may not be warranted. In order that these test modalities can still be offered by labs, a potential area of growth in the future may be the establishment of third-party commercial measurement technology centers. Such centers would provide remote-site high-tech measurement capability. These centers would be capable of assuming the risk associated with rapidly changing technology by making high-tech services for skin measurement available to many different industries through selected cooperating contract clinical test facilities.

REFERENCES

1. Stehlin, D., Erasing wrinkles: Easier said than done, FDA Consumers, July-August 20-22, 1987.
2. Jackson, E. M., How to choose a contract laboratory: Utilizing a laboratory clearance procedure, J. Toxicol., 3(2): 83-92, 1984.
3. Sampson, M. C., Career opportunities in industrial clinical research. In The Clinical Research Process in the Pharmaceutical Industry (G. M. Mataren, ed.), 1984, pp. 479-507.
4. Arnold, J., and Haydan, D. M., Role of contract research organizations. In The Clinical Research Process in the Pharmaceutical Industry (G. M. Mataren, ed.), 1984, pp. 379-393.
5. Merker, P. C., Good laboratory practices (GLPs) and good clinical practices (GCPs): Beneficial impact on safety testing, J. Toxicol., 3(2):83-92, 1984.

2

Irritation and Sensitization

Edward M. Jackson

Noxell Corporation, Hunt Valley, Maryland

I. INTRODUCTION

Irritation and sensitization are the most common causes of inflammation in a spectrum of inflammatory responses which includes light-mediated irritation (phototoxicity or photoirritation), light-mediated sensitization (photoallergenicity), and acnegenicity. Of these two causes of inflammation, the frequency of irritation is far greater than the frequency of sensitization and all other causes of inflammation combined.

In this chapter we begin with the physiology and biochemistry of inflammation. There follows a detailed discussion of both objective and subjective irritation as a cause of inflammation, and the tests used to assess the irritation potential of cosmetic products. Sensitization as a cause of inflammation is discussed next, along with the tests to determine the sensitization potential of cosmetic products. Sensitization as a cause of inflammation is approached both from the prognostic and diagnostic viewpoints. Before concluding this chapter, the legal and professional responsibilities of the scientists and clinicians who substantiate the safety of cosmetic products in tests on human volunteers are reviewed.

II. INFLAMMATION

Inflammation is the body's general distress response to biological, physical, or chemical causes of irritation or sensitization. This definition is broad enough to apply to the topical as well as the systemic inflammatory processes. This portion of the chapter will concentrate on the events involved in cutaneous inflammation. A further focus in this chapter will be on chemicals causing inflammation either by irritation or sensitization [1].

Clinically, inflammation is still defined by the four classic and characteristic signs of erythema, edema, pain, and/or heat. Our measurement of the degree of erythema or edema is still crude. Despite this, discussion of the degree of erythema or edema persists in the scientific and clinical literature. Professionals are more likely to think in terms of none, more, and marked when discussing these two clinical signs of inflammation. Measuring pain or heat on the other hand is even more complex. Even crude estimates are made more complex and less definable owing to the subjective aspects of pain or heat versus the more objective inflammatory signs of erythema and edema [1].

A. Cellular Events

Figure 2.1 is an abbreviated diagram of the cell containing only those subcellular structures which have been demonstrated to play some role in the inflammatory process at the cellular level.

Inflammation begins at the cellular level. The cell membrane is noted for its selective permeability and molecular attachment sites. While little specific research has actually been able to demonstrate specific events at the site of this cell membrane in inflammation, it should be obvious that chemicals which transiently or permanently alter the selective permeability of the cell membrane or the binding capacity of the molecular attachment sites will directly influence the inflammatory process at the cellular level. For example, surfactants alter the permeability of the cell membrane [2,3]. The ability of the cell to take in substances through vacuolization [4] results in the production of other key cellular structures, the vacuoles. The contents of such vacuoles can be dangerous to the survival of the cell. Lysosomes, however, are designed to deal with this situation. Vacuolization is a two-way street. The cell uses this process to concentrate waste materials it needs to discard from products it makes, to move them to the membrane, and then to expel these wastes outside the cell. Both the cell membrane and the

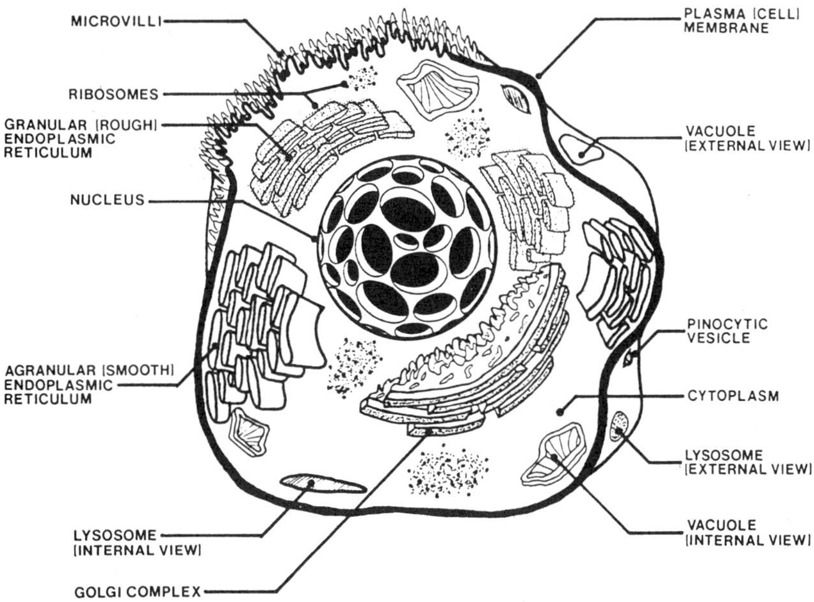

FIGURE 2.1 Key organelles in the response of the cell to inflammation [1].

process of vacuolization are, therefore, related, since a cellular vacuole is encircled by a pinch of cell membrane.

The nucleus has been shown to be operative in the process of inflammation. Specifically, deoxyribonucleic acid (DNA) synthesis has been shown to increase during exposures to chemical irritants [5]. However, further research into the activities of the cell nucleus under specific stresses by chemical irritants or sensitizers needs to be undertaken. Alterations of the cell type at specific points may also be a fertile avenue of further inquiry.

The endoplasmic reticulum (ER) has also been shown to be affected by inflammation. Changes in either resting (smooth) or active (rough) ER can both occur during the process of inflammation. Ribosomes are, of course, the ribonucleic acid (RNA)-protein conglomerates that give the ER its rough appearance indicating protein synthesis. Therefore, the concentrations of polyribosomes and the ribosomal subunits could be indicators of effects of chemicals which activate the inflammatory process [6].

However, the cellular organelle which we know most about in the process of inflammation is the lysosome. This important and interesting cellular organelle was discovered and elucidated by de Duve [7]. The lysosome contains all the hydrolytic enzymes capable of digesting all the parts of the cell. Its role in digesting and neutralizing the contents of vacuoles has already been mentioned. Even though the lysosome has been known since 1955, it was not until 1975 that the lysosome was suggested as possibly being involved in inflamed tissue and leukocyte chemotaxis [8]. Middleton [9] demonstrated the usefulness of enzyme release as a measure of cellular toxicity, which would, of course, include the cellular effects of inflammation. Hirschhorn had already tied disruption of the cell membrane to lysosomal membrane rupture, and further described selected lysosomal rupture after certain chemicals were concentrated in the lysosomes through their contact with vacuoles [10]. Selective lysosomal disruption from exposure to photosensitizers has also been demonstrated [11]. One may assume that the compounds which stabilize either the cellular or lysosomal membranes would, therefore, be capable of preventing or at least reducing the effects of cellular inflammation. There may even be relationships between the activities of the lysosome and inflammation and certain disease states such as lupus, atopic dermatitis, and pemphigus vulgaris [8]. More recently, Gibson has shown the selective effect of certain surfactants on lysosomal enzyme release and made very interesting correlations between visible erythema and edema, cellular lysosomal enzyme release, and a histological increase in the numbers of leukocytes in edematous tissue [12,13].

The lysosome then continues to be the single most important cellular organelle in the process of cutaneous inflammation.

B. Biochemical Events

Figure 2.2 outlines the basic biochemical pathway for the conversion of phospholipids to arachidonic acid, which then can be converted to precursors of the prostaglandins (PGs), hydroperoxyeicosatetraenoic acid (HETE) and the leukotrienes. Figure 2.2 is simplified by omitting the actual chemical structures and by listing only the significant steps in the pathway which have been a focus of major research efforts. What is interesting about Fig. 2.2 is that it also summarizes the incredible amount of biochemical research which has been reviewed [14-16].

Phospholipids are the major raw material and starting point for the arachidonic acid pathway and are easily supplied to the

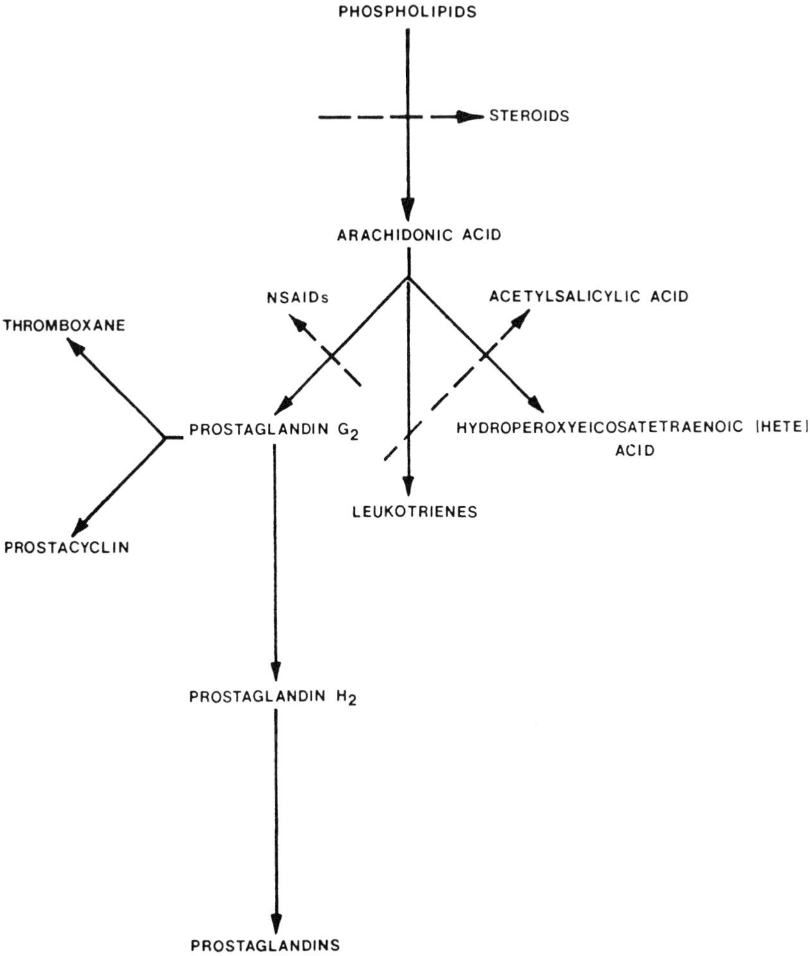

FIGURE 2.2 Arachidonic acid pathway [1].

cell through the diet. We now know that chemical irritants increase phospholipid synthesis [5] and trigger our biochemical response to irritation by stoking, as it were, the production of the various molecular species important in responding to an insult. Phospholipid lipase is the enzyme responsible for the conversion of phospholipids to arachidonic acid. Steroids inhibit the conversion of phospholipids to arachidonic acid. From a clinical standpoint, chemical irritation can be reduced by the topical application of hydrocortisone creams or lotions, which are available either as prescription drugs or over-the-counter products. However, hydrocortisone creams may not be the only anti-inflammatory topical treatments. Simpler treatments like petrolatum may also be used, since petrolatum has been shown to reduce the production of certain prostaglandins in animal skin [17]. It is interesting to note that petrolatum is a classic vehicle and a typical negative control in testing in diagnostic patch testing. It has been assumed to be nonreactive with the skin. This may not actually be the case.

Arachidonic acid is resident in the cellular membrane where it is the source of several major biochemical pathways. If arachidonic acid is acted on by the enzyme cyclooxygenase, prostaglandin G_2 is generated, which itself can be converted into thromboxane or prostacyclin or prostaglandin H_2, the latter then generating the other members of the prostaglandin family. The nonsteroidal anti-inflammatory drugs (NSAIDs) inhibit the conversion of arachidonic acid to prostaglandin G_2. But another inhibitor of prostaglandin synthesis has been demonstrated, i.e., nutmeg oil [18]. Eugenol, a major component of nutmeg oil, was further shown to inhibit thromboxane formation by the same researchers.

Thromboxane and prostacyclin are biochemical products of prostaglandin G_2 which produce opposite effects. Thromboxane stimulates platelet aggregation in the clotting process and is a vasoconstrictor, countering the vasodilatative effects from materials such as histamine. Prostacyclin, on the other hand, inhibits the platelet aggregation process and vasoconstriction.

Prostaglandin G_2 can further be biochemically transformed to prostaglandin H_2, which is the beginning of all other prostaglandin species, including PGE_2 and PGI_2, which are the key chemical species in the process of inflammation.

The study of prostaglandins, of course, is a fascinating chapter in the history of biochemistry; named as they were because they were first discovered in human semen. The whole family of prostaglandins is characterized by extreme potency. Also, they were the first nonprotein chemical mediators discovered, since they are fatty acids. Until the discovery of prostaglandins hormones, which are protein, had been the only known biochemical mediators.

As far as the process of inflammation is concerned, PGE_2 is critical to the production of vasodilatation (and, therefore, erythema), pain, and edema [16].

Changing now to the second major arachidonic acid pathway, the enzyme 12-lipoxygenase results in the formation of hydroperoxyeicosatetraenoic acid (HETE) from arachidonic acid. 5-Lipoxygenase catalyzes the production of the eicosanoid family known as leukotrienes. Specific leukotrienes produce a multitude of proinflammatory effects which include bronchoconstriction; increased vascular permeability in postcapillary venules; the production of mucus, which is chemotactic to leukocytes; and the production of superoxides from neutrophils [19]. Leukotrienes have also been implicated in chronic dermatoses, including psoriasis [20]. Certain doses of acetylsalicyclic acid (ASA) have been demonstrated to inhibit both of these pathways.

The continued elucidation of the arachidonic acid pathway will lead to a more complete understanding of inflammation, and result in the successful treatment of certain disease states and topical products which are formulated to be used with less potential for cutaneous reactions.

III. IRRITATION

Irritants are chemical, physical, or biological agents which produce inflammation that can be objectively observed (erythema, edema, and heat), or subjectively described (pain, stinging, burning, or itching). The process of chemical irritation has already been reviewed and adequately described [5,21]. In summary, the result of chemically insulting the skin is the release of histamine in the irritated area by the mast cells. This potent vasodilator produces the visible erythema and increased vascular permeability (resulting in the leaking of fluid, therefore, edema) allowing cells to migrate to the irritated area. Chemotaxis of the polymorphonucleocytes (PMNs) results. The PMNs marginate along the capillary walls and diapedese to the irritated area of the skin.

Although the inflammatory response to an irritant occurs immediately and only at the site of the insult, it is modified by the strength of the irritant. For example, the skin can be exposed to a chemical irritant in its concentrated or dilute state. The duration of the exposure to the irritant can also modify how strong the inflammatory response will be. The concentration of the irritant and length of the exposure to the irritant then are modifiers of the immediate inflammation produced at the site of contact.

Of all the causes of inflammation, irritants far outnumber the skin's responses to other inflammatory causes such as sensitizers, photoirritants, or photoallergens. This means we are exposed to irritants far more than any other type of inflammatory cause. One reason for this is that, in a sense, irritation tells us something about how we are all the same, whereas sensitization tells us about how we are all different [22]. All of us respond to an irritant such as a solvent or an acid, but only some of us will respond to a given sensitizer like the poison ivy allergen (5-pentadecylcatechol). The irritant response is universal, the sensitization response is individual and specific.

A. Objective Irritation

The clinical assessment of irritation potential can be made by patch testing (prophetic patch test, repeat insult patch test, 21-day cumulative irritancy test), or controlled-use testing.

In a patch test, the test material is applied to a Band-Aid-like material, or patch, and pressed onto the panelist's skin (usually the back or forearm). After 24-48 hr, the patch is removed, the skin allowed to equilibrate for approximately 30 min, and the skin scored for erythema and edema. The International Contact Dermatitis Research Group (ICDRG) scoring scale of 0 (no reaction), 1+ (erythema), 2+ (erythema and edema), and 3+ (marked erythema and edema) is routinely used [23]. The reading for irritation can thus be made immediately. The difference between the prophetic, repeat insult, and 21-day cumulative irritancy test is simply the number of applications of test material (or patches) made.

The prophetic patch test [24-26] was developed during the 1930s and altered the original use of the patch test as a diagnostic tool. It changed it from a diagnostic tool to a prognostic test. This is significant because this permitted premarket safety testing by screening for both sensitizers and irritants. It has been further modified today, along with the repeat insult patch test, to screen for photosensitizers as well.

The prophetic patch test is a screening test for rank irritants, strong sensitizers, and photosensitizers. The repeat insult patch test is a predictor of low-level irritants and sensitizers. Further, the repeat insult patch test can be modified to distinguish between photoirritants and photoallergens in photosensitization reactions.

The repeat insult patch test [27,28] was first published by the Shelanskis and later by Draize. But no research has ever been published by both originators of the test, and the so-called Draize-Shelanski Repeat Insult Patch Test is a misnomer [29].

The 21-day cumulative irritancy test [30-33] is the true test for determining irritation potential. Originated by Lanman [30] and modified by Phillips [31], two improvements have been proposed to make this test even more effective in sorting through ingredients and products of low irritation potential. Berger [32] proposed shortening the test to 14 days from 21 days, since 95% of the data is effectively generated in that time. This also eliminates the skin fatigue phenomenon which can be so troublesome in interpreting results toward the end of this type of test. Carabello [33] has proposed analysis of results by a frequency index (FI), which is the mathematical version of graphing the cumulative irritation scores. Using this approach to previously generated 21-day cumulative irritancy test results can yield more information than was previously thought to be contained in the results from this test.

Patch testing, whether prophetic patch testing, repeat insult patch testing, or 21-day cumulative irritancy testing, cannot be used alone to determine the irritation potential of an ingredient or product. It must be coupled with a controlled-use test over a defined time period, like 28 days [34-36], which provides for the actual use of the product in its final package on the body part the product is designed for. Exaggerated exposures can also be utilized in the protocol. It is a fact that a product can pass patch testing, yet fail controlled-use testing because patch testing on the skin of the back or forearm can be quite different from use testing the same product on the periorbital or perioral area for example. Further, patch testing produces only objective irritation responses, whereas controlled-use testing yields both objective and subjective irritation responses.

The chamber test [37] was an attempt to determine the irritation potential of a mild surfactant, a detergent, or a soap solution. More would be said about the chamber test except that one of the originators has demonstrated that actual in-use testing (controlled-use testing) is a better determination of the irritation potential of such cleansers than the chamber test [38]. Additional support for this conclusion is contained in research from a study conducted on an over-the-counter 10% benzoyl peroxide product [38]. In that study, it was clearly demonstrated that the use of an emollient cleanser rather than soap aided in the reduction of open comedones, papules, and pustules, and was equivalent in reducing closed comedones during the first 28 days of an 84-day washing regimen versus all other soap and benzoyl peroxide wash products. The single variable in the washing regimen test [39] was the cleansing agent.

B. Subjective Irritation

Subjective irritation testing attempts to assess the phenomenon of stinging, burning, and itching [40]. These reactions are true irritation reactions and must be considered when attempting to determine the irritation potential of a product. We are just beginning to understand the relevancy of this class of irritation response. The one common denominator to both objective irritation (which is visible as erythema and palpable as edema) and subjective irritation (which is felt as burning, stinging, or itching) is the fact that both are a result of the percutaneous absorption of the product [41]. In the former case, vasodilatation of the vasculature in the papillary dermis produces erythema and edema. The nerve supply of the skin is located almost exclusively in the dermis, accounting for the responses of pain, heat, and the stinging, burning, and itching sensations from subjective irritation as well. When these responses occur percutaneous absorption has occurred through both the stratum corneum of the epidermis and the epidermal-dermal junction.

In summary, the use of both a properly designed patch test and a controlled-use test can substantiate the safety of a topical product by properly assessing its irritation potential prior to introduction into the marketplace. Assessing the irritation potential of a product is at least as important as assessing the sensitization potential because of the high frequency of irritation as a cause of inflammation.

IV. SENSITIZATION

We have of necessity discussed several aspects of sensitization in the section on irritation. For example, sensitization as a cause of inflammation is not as frequent a reaction as is irritation. This is because sensitization tells us how we are different from one another. Different people react differently to various sensitizers.

Yet it is an interesting paradox that more is written about the process of the inflammatory response to sensitizers, both biochemically and at the cellular level, than is written about irritation reactions. Whole journals, monthly columns in periodicals, and innumerable research articles are devoted to sensitization. Not so for irritation.

The inflammation resulting from exposure to chemicals known as sensitizers or allergens is a much different cellular process than the inflammation resulting from the exposure to irritants. Sensitization is a response of the immune system, but it is *not* an antigen-antibody reaction. This early but erroneous understanding of the sensitization reaction was actually written into the 1960 Federal Hazardous Substances Act regulations governing the testing of household products [45]. These regulations are currently administered by the Consumer Product Safety Commission. After 25 years, this erroneous and outdated science was set straight in 1986 [46], but only after much confusion in the regulatory literature. Once again, we have learned that science should never be written into law precisely because of the organic and evolutionary nature of the process of science.

Skin sensitization is a delayed-type humoral immune response mediated by the T cell [47,48]. The incomplete allergen or hapten complexes with any of a number of proteins in the skin. The Langerhans cell in the zone of the stratum germinativum interacts with this allergen, forming a hapten-protein complex and migrates to the thymus gland. Like a professor returning to a university after a sabbatical leave, the Langerhans cell then "teaches" naive or uneducated thymus T cells about the allergen. The education of these naive T cells is complete when they leave the thymus as sensitized T cells, and on contacting the allergen, liberate lymphokines. Lymphokines are chemotactic agents which attract other leukocytes to the site [47], and appear to raise the temperature of the area under stress in an attempt to rid it of the material producing the inflammation. Unfortunately, this temperature effect is only effective against certain biological agents, such as bacterial or protozoan infectious agents.

We previously discussed cellular involvement in inflammation at the epidermal cell level. Mast cells, as producers of histamine, have been studied in response to surfactants which alter the mast cell membrane. Histamines are the powerful chemical mediators of the allergic response [50-52] and are released in response to various insults, including chemical sensitizers.

The prognostic or diagnostic use of the patch test is one of two ways to determine the sensitization potential of a product. Historically, the patch test was first used in 1847 by a chemist named Staedler to demonstrate sensitization to the allergen Balsam of Peru. In 1895, Josef Jadassohn, one of the nineteenth century's greatest dermatologists, who lived well into the twentieth century, demonstrated skin sensitization to mercury in an ointment prescribed for a patient of his suffering from *Pediculus pubis* (pubic lice, or crabs) in the pubic region [53-55]. The patch technique was fully developed by Jadassohn as a diagnostic tool. Shortly thereafter, an entymologist, Fabre, demonstrated sensitization to a species of caterpillar hair through a method of application of the hair on a blotting paper square [55]. Two things are clear from the history of patch testing. First, patch testing was originally proposed as a diagnostic tool to determine a reaction to a drug. Secondly, patch testing was used to determine sensitization as the drug reaction, not irritation.

To determine sensitization in patch testing, a 2-week rest period is utilized after induction (single or multiple initial exposures to the test material). During the induction, the T cells are activated and circulate throughout the body. When the skin is insulted again through the reapplication of a patch with the same test material the T cells liberate lymphokines, producing the delayed-type humoral response described above. After removing the challenge patch, readings should be made at 24, 48, 72, and even 96 hr to ensure a negative or a positive sensitization response.

One final test method deserves mention in this section on sensitization tests. It is the maximization test procedure [42-44]. This prognostic and diagnostic tool is based on chemically eliminating the barrier properties of the stratum corneum to permit ready access of a potential allergen to the Langerhans cells. In this way, weak sensitizers can be detected. However, except in those cases where detection of weak sensitizers is required, the patch test coupled with the use test should be preferred.

To determine sensitization in a controlled-use test, the same 2-week rest period is utilized after the use period (induction), after which a challenge patch is applied and read under the same circumstances and at the same time or intervals as in patch testing.

In concluding this section on the cellular events of inflammation, the subjects of anti-irritancy and quenching sensitizers (allergens) also have a place. The reason is that what we know about anti-inflammation and quenching may aid our understanding of cellular inflammation, enabling us to either control or eliminate it or formulate around it. Goldemberg has extensively reviewed the subject of anti-irritancy [56-59]. The reduction in the cutaneous or ocular irritation potential of an ingredient has been achieved in certain formulations where chemical complexations of the irritant have occurred, or the reactive cell sites have been blocked, thereby reducing to some degree the irritant effect of the neat chemical. Also, as one example of a specific anti-inflammatory molecule, superoxide dismutase (SOD) has been used effectively to reduce the inflammation from arthritis.

Quenching known sensitizers or allergens is also a possibility in certain defined situations [60-64]. Our understanding of this area above all others needs to be expanded to aid in formulating chemicals which have a sensitization potential in their neat state into products which have little or no sensitization potential.

V. LEGAL AND PROFESSIONAL RESPONSIBILITIES

The use of human subjects to determine the irritation or sensitization potential of a product is a serious ethical burden on both the scientists and physicians conducting these tests. In this regard, both the scientists and the physicians are bound by their respective profession's code of ethics to determine irritation or sensitization potential without altering, endangering, or compromising the health of the panelists in the test.

Added to these ethical responsibilities are the regulatory requirements which are subsumed under the general heading of the good clinical practice (GCP) regulations. While it is true that these GCP regulations are not complete at this time, there are several parts of the proposed GCPs which have already been proposed.

The obligations of the clinical investigators, sponsors, and monitors were proposed in 1977 [64] and 1978 [65]. Standards for institutional review boards were also proposed in 1978 [66]. The protection of human subjects by the use of an informed consent and release form was proposed in 1981 [67].

Reading through these documents, it is clear that the U.S. Food and Drug Administration considers testing in human subjects to be a serious regulatory responsibility of both the scientists and physicians involved in this type of testing. Although each item in these proposed regulations is not per se applicable to the testing of cosmetic products, many companies employ these or similar practices for the protection of their test subjects.

Finally, the litigious society we have evolved places a heavy legal burden not only on the company sponsoring the test as well as its monitors, but upon the clinical investigators themselves.

Companies must be prepared for law suits based on any product-related adverse effects resulting from testing their products for irritation or sensitization potential. As one example, many companies employing contract laboratories now legally indemnify the laboratories, since the laboratories cannot economically withstand even a single suit.

In summary, there can be no compromise on the professional, regulatory, and legal responsibilities of scientists and physicians using human subjects to determine the irritation or sensitization potential of a product.

CONCLUSIONS

The irritation and sensitization potential of cosmetic products must be determined prior to introduction into the marketplace. Of these two causes of inflammation, irritation is clearly the more frequent reaction from the application of topical products. However, sensitization is certainly the more publicized of the two types of reactions, and because of the involvement of the immune system, draws much more attention in research articles and articles describing reactions from cosmetic products.

We now understand that irritation is a combination of objectively observed phenomena and subjectively described responses to the application of topical products. Subjective irritation testing needs to be developed further to be able to eliminate any subjective irritation potential from the use of topical products.

Our current understanding of the biochemistry and cell physiology of inflammation points to several conclusions. Among these are the realization that the major pieces of the cellular and molecular puzzle of inflammation are now in place. This juncture in our continuing effort to understand, control, and treat information is a critical one which must be appreciated to effectively continue our research and application of this knowledge.

We now need to bridge the gap between our understanding of the topical and systemic inflammatory processes. There are common denominators here which scientists in both areas can benefit from sharing. For example, topical inflammation can result from exposure to irritants, sensitizers, or photosensitizers. Systemic inflammation on the other hand is principally caused by irritants.

At present, the patch test and use test are adequate to determine the irritation and/or sensitization potential of topical products such as cosmetics. However, we all recognize that even these tests must become more sensitive to detecting these causes of inflammation and new tests must be developed.

The tests to determine the irritation or sensitization potential of an ingredient or product are still at a crude stage of our understanding of how to test for these causes of inflammation. In a sense, the irritation or sensitization potential of a product is determined after the fact and not while inflammation is actually initiated or developing. After inflammation actually occurs, it is diagnosed simply by observing the involved skin to raised or red.

What can we expect from irritation and sensitization testing in the future?

First, we may see a change from the classic signs of inflammation (erythema, edema, pain, and heat) to a combined set of both objective signs of inflammation (erythema, edema, and heat) in conjunction with a set of subjective signs of inflammation (pain, itching, stinging, and burning). Perhaps then we will be better able to recognize both objective and subjective irritation as simply different ways for the inflammatory response to manifest itself.

Second, we may see the development of better and more sensitive diagnostic tests with more refined scoring scales and better ways to both read and interpret the results. Since we are still using 0, 1+, 2+, and 3+ to denote no erythema, erythema only, erythema and edema, and marked erythema and edema, respectively, the scoring scales may still be objective but quite different. Perhaps these scoring scales will be based on the percutaneous absorption or permeability of a topical product, its somatosensory effects, photopletysmogryphic techniques (which measure the flux in the dermal capillary bed), and even thermography.

Third, the rapidly expanding field of percutaneous absorption itself could lead to a more complete understanding of both the skin and its function of a semipermeable membrane. Knowing more about percutaneous absorption will teach us more about irritation and sensitization as an inflammatory response.

Finally, the subject of subclinical irritation must be raised. Subclinical irritation is an inflammatory response which cannot be observed or felt but is histologically verifiable. We know that many products produce subclinical irritation. What we have now come to realize is that subclinical irritation can be a stimulus for increased epidermal cell turnover rate, hypersensitivity, and even acne flare ups. Reducing or eliminating the potential for subclinical irritation from the use of topical products is currently not within our grasp. But capitalizing on it when it does occur may be.

In summary, we have adequate tests to determine the irritation and sensitization potential of the products we are currently formulating, but we must actively work to develop better and more discriminating tests which will result in even better cosmetic products.

REFERENCES

1. Jackson, E. M., Cellular and molecular events of inflammation, J. Toxicol. Cut. Ocular Toxicol., 3(4):347, 1984.
2. Prottey, C., and Ferguson, T. F. M., The effect of surfactants upon rat peritoneal mast cells in vitro, Food Chem. Toxicol., 14:425, 1976.
3. Ferguson, T. F. M., and Prottey, C., The effect of surfactants upon the mammalian cells in vitro, Food Chem. Toxicol., 14:432, 1976.
4. Kolata, G., "The" magic in magic bullets, Science, 222: 310, 1983.
5. Prottey, C., The molecular basis of irritation. In Cosmetic Science, vol. 1 (M. M. Breuer, ed.), Academic Press, New York, 1978.
6. Brubaker, P. E., Alexander, L. T., and Jackson, E. M., Conservation of Ribosomes During the Cell Cycle, presented at the Annual Meeting of the American Society of Cell Biology, New Orleans, November, 1971.
7. de Duve, C., Pressman, B. C., Gianetto, R., Wattiaux, P., and Appelmans, F., Tissue fractionation studies. 6. Intracellular distribution patterns of enzymes in rat liver tissue, Biochem. J., 60:604, 1955.
8. Lazarus, G. S., Hatcher, V. B., and Levine, N., Lysosomes and the skin: A review, J. Invest. Dermatol., 65: 259, 1975.
9. Middleton, Evaluation of cellular injury in skin utilizing enzyme activities in suction blister fluid, J. Invest. Dermatol., 74:219, 1980.
10. Hirschhorn, R., Lysosomal mechanisms in the inflammation process. In The Inflammation Process (B. W. Zweifach, ed.), Academic Press, New York, 1974.

11. Allison, A. E., Magnus, I. A., and Young, M. R., Role of lysosomes and cell membranes in photosensitization, Nature, 209:874, 1976.

12. Gibson, W. T., and Teall, M. R., Interactions of C12 surfactants with the skin: Studies in enzyme release and percutaneous absorption in vitro, Food Chem. Toxicol., 21(5):581, 1983.

13. Gibson, W. T., and Teall, M. R., Interactions of C12 surfactants with the skin: Studies in enzyme release and percutaneous absorption in vitro, Food Chem. Toxicol., 21(5):587, 1983.

14. Kuehl, F. A., Jr., and Egan, R. W., Prostaglandin, arachidonic acid and inflammation, Science, 210:978, 1980.

15. Nelson, N. A., Kelley, R. C., and Johnson, R. A., Prostaglandins and the arachidonic acid cascade: A special report, Chem. Eng. News, August 16, 1982, pp. 30-44.

16. Goldyne, M. E., Eicosanoids and human skin, Prog. Dermatol., 20:(1), March 1986.

17. Pennys, N. S., Eaglstein, W., and Ziboh, V., Petrolatum: Interference with the oxidation of arachidonic acid, Br. J. Dermatol., 301:257, 1980.

18. Rasheed, A., Laeckerman, G., Totte, J., Vlietinck, A. J., and Herman, A. G., Eugenol and prostaglandin biosynthesis, N. Engl. J. Med., 301(1):50, 1984.

19. Samuelson, B., Leukotrienes: Mediators of immediate hypersensitivity reactions to inflammation, Science, 220:568, 1983.

20. Voorhees, J. J., Leukotrienes and other arachidonic acid products in the pathogenesis and therapy of psoriasis and other dermatoses, Arch. Dermatol., 119(7):541, 1983.

21. Guyton, A. C., Textbook of Medical Physiology, 4th ed., Saunders, Philadelphia, 1971.

22. Jackson, E. M., The difference between irritation and sensitization, J. Toxicol. Cut. Ocular Toxicol., 4(1):1, 1985.

23. International Contact Dermatitis Research Group (ICDRG), Terminology of Contact Dermatitis ACTA Dermatovener, 50:287, 1970.

24. Schwartz, L., and Peck, S. M., The patch test and contact dermatitis, Public Health Rep., 59:2, 1944.

25. Brunner, M. J., and Smiljanic, A., Procedure for the evaluation of the skin sensitizing power of new materials, Arch. Dermatol., 66:703, 1952.

26. Traub, E. F., Tusing, T. W., and Spoor, H. J., Evaluating dermal sensitivity, Arch. Dermatol., 69:399, 1954.

27. Shelanski, H. A., and Shelanski, M. V., A new technique of human patch tests. In Proceedings of the Scientific Section of the Toilet Goods Association (now called the Cosmetic, Toiletry and Fragrance Association), Washington, D.C., 19:46, 1953.

28. Draize, J. H., Dermal toxicology, Food Drug Cosmet. J. Law, 10:722, 1955.

29. Shelanski, M. V., The patch test. I. The history and review of methods, capabilities and inherent limitations, J. Toxicol. Cut. Ocular Toxicol., 1(2):91, 1982.

30. Lanman, B. M., Zelvers, W. V., and Howard, C. S., The role of human patch testing in the product development program. In Proceedings of the Joint Conference of Cosmetic Sciences, the Toilet Goods Association (now called the Cosmetic, Toiletry and Fragrance Association), Washington, D.C., 1968.

31. Philips, L., Steinberg, M., Maibach, H. I., and Akers, W. A., Comparison of rabbit and human skin responses to certain irritants, Toxicol. Appl. Pharmacol., 21:639, 1972.

32. Berger, R. S., and Bowman, J. P., Early appraisal of the 21 day cumulative irritation test in man, J. Toxicol. Cut. Ocular Toxicol., 1(2):109, 1982.

33. Carabello, F. J., The design and interpretation of human skin irritation studies, J. Toxicol. Cut. Ocular Toxicol., 4(2):61, 1985.

34. Reiger, M. M., and Battista, G. W., Some experiments in the safety testing of cosmetics, J. Soc. Cosmet. Chem., 15:161, 1964.

35. Idson, B., Topical toxicity testing, J. Pharm. Sci., 57(1): 1, 1968.

36. Jackson, E. M., and Robillard, N. F., The controlled use test in a cosmetic product safety substantiation program, Food Chem. Toxicol., 1(2):117, 1982.

37. Frosch, P. J., and Kligman, A. M., The Chamber-scarification test of irritancy, Contact Dermatitis, 2:314-324, 1976.

38. Mills, O. H., Swinyer, L. J., and Kligman, A. M., Assessment of Irritancy of Cleansers. Presented as a scientific exhibit at the 43rd Annual Meeting of the American Academy of Dermatology, Washington, D.C., December 1-6, 1984.

39. Jackson, E. M., The Effect of Cleansing Regimens on the Success of Acne Therapy Using 10% Benzoyl Peroxide Lotion. Presented at the 4th Annual Summer Session of the American Academy of Dermatology, Boston, June 26, 1986.

40. Frosch, P. J., and Kligman, A. M., A method for appraising the stinging capacity of topically applied substances, J. Soc. Cosmet. Chem., 28:197, 1977.

41. Jackson, E. M., The toxicological aspects of percutaneous absorption data. In Percutaneous Absorption: Fundamentals and Applications (J. L. Zatz, ed.), Aliured Publishing, Chicago, 1990.

42. Kligman, A. M., The identification of contact allergens by human assay. I. A critique of standard methods, J. Invest. Dermatol., 47(5):369, 1966.

43. Kligman, A. M., The identification of contact allergens by human assay. II. Factors influencing the induction and measurement of allergic contact dermatitis, J. Invest. Dermatol., 47(5):375, 1966.

44. Kligman, A. M., The identification of contact allergens by human assay. III. The maximization test: A procedure for screening and rating contact allergens, J. Invest. Dermatol., 47(5):393, 1966.

45. Code of Federal Regulations, 2500.3(b)(9) and 1500.3(c)-(5)b.

46. Hazardous Substances, Supplementary Definition of Strong Sensitizer, Fed. Reg., 51(157):29094, August 14, 1986.

47. Coombs, R. R. A., and Gell, P. G., Classification of allergic reactions responsible for clinical hypersensitivity and disease. In Clinical Aspects of Immunology (P. G. H. Gell and R. R. A. Coombs, eds.), Oxford University Press, Blackwell, England, 1975.

48. Benezra, C., Irritant and allergic properties of chemically defined fragrance substances, Cosmet. Technol., 20-26, February, 1980.

49. Snydermann, R., and Goetzel, E. J., Molecular and cellular mechanisms of leukocyte chemotaxis, Science, 213:830, 1981.

50. Pepys, J., and Edwards, A. M. (eds.), The Mast Cell: Its Role in Health and Disease, University Park Press, Baltimore, Maryland, 1979.

51. Kalmer, M. A., The mast cell—A fascinating riddle, N. Engl. J. Med., 301(9):498, 1979.

52. Ishizaka, T., Analysis of triggering events in mast cells for immunoglobin E—Mediated histamine release, J. Allergy Clin. Immunol., 67(1):90, 1981.

53. Jadassohn, J., A contribution to the study of dermatoses produced by drugs, Verh. Dtsch. Dermatol. Ges., 103-129, 1895.

54. Jadassohn, J. (1863-1936) Dermatologist, an editorial in JAMA, 207(8):1511, February 24, 1969.

55. Foussereau, J., History of epicutaneous testing: The blotting paper and other methods, Contact Dermatitis, 11:219, 1984.

56. Goldemberg, R. L., Use of anti-irritants in cosmetic formulating, J. Soc. Cosmet. Chem., 16:317, 1965.

57. Goldemberg, R. L., and Safrin, L., Reduction of topical irritation, J. Soc. Cosmet. Chem., 28:667, 1977.

58. Goldemberg, R. L., Anti-irritants, J. Soc. Cosmet. Chem., 30:415, 1979.

59. Suskind, R., and Majuti, V., Occupational and environmental allergic problems of the skin, J. Dermatol., 3:3, 1976.

60. Opdyke, D. L. J., Inhibition of sensitization reactions induced by certain aldehydes, Food Chem. Toxicol., 14:197, 1976.

61. Goldemberg, R. L., Quenching sensitization potential, Skin Allergy News, 7(9), September 1976.

62. Goldemberg, R. L., Perfume sensitization, Drug Cosmet. Ind., 119:28, April 1976.

63. Fisher, A. A., and Dooms-Goosens, A., The effect of perfume aging on the allergy of individual perfume ingredients, Contact Dermatitis, 2:155, 1976.

64. Clinical Investigations, Obligations of Sponsors and Monitors, Fed. Reg., 42(187):49612, September 27, 1977.

65. Obligations of Clinical Investigators of Regulated Articles, Fed. Reg., 43(153):35210, August 8, 1978.

66. Standards for Institutional Review Boards for Clinical Investigators, Fed. Reg., 43(153):35186, August 8, 1978.

67. Protection of Human Subjects, Informed Consent, Fed. Reg., 46(17):8942, January 27, 1981.

3

Phototoxicity and Photoallergy

Ward L. Billhimer

Hill Top Research, Inc., Cincinnati, Ohio

I. INTRODUCTION

The skin is a unique organ in that it interfaces with the environment, and among other functions, shields the body from the effects of the ultraviolet component of sunlight. Topically applied materials, such as cosmetics and drugs, can be affected by ultraviolet radiation such that the health of the skin is seriously compromised. In fact, over the last several decades, adverse skin reactions resulting from the interaction of sunlight and chemicals have been widespread. Such diverse materials as coal tar derivatives, drugs, topical medicaments, fragrances, dyes, numerous cosmetics and toiletries, sunscreens, soaps, detergents, and optical bleaches have been involved.

During World War II an epidemic of light-dependent rashes associated with the antibiotic sulfonamide developed in 361 military patients in North Africa and Italy [1]. Almost 20 years later, another epidemic of much larger proportion resulted from the introduction of certain topical antimicrobials. Between 1960 and 1961, nearly 10,000 patients in England presented with light-induced skin allergies to antimicrobial soaps containing the halogenated salicylanilide tetrachlorosalicylanilide (TSCA) [2,3]. In efforts to replace TSCA, several other antimicrobials, including bithionol, hexachlorophene, dichlorophene, the carbanilides, bis[2-hydroxy-5-chlorophenyl]sulfide, chlorosalicylamide, and chloro-2-phenylphenol, were introduced into widely distributed products such as deodorant soaps. By 1970, even these materials had been associated with such reactions, but to a much lesser extent than TSCA [4,5].

In the late 1970s, two fragrance materials, 6-methylcoumarin (6-MC) and musk ambrette, provoked light-dependent dermatitis [6-8]. At that time, 6-MC was used in a variety of cosmetics and toiletries, including some sunscreen and suntan products. Likewise, musk ambrette could be found in many soaps, perfume formulations, detergents, creams, lotions, dentifrices, and other personal care products.

Considering their widespread use and photoactive nature, sunscreen agents have generally exhibited a low incidence of photosensitization reactions [9]. Para-aminobenzoic acid esters are most often reported to cause photosensitivity; however, other sunscreen agents, such as benzophenones, cinnamates, and digalloyltrioleate, have also been implicated [10-13].

Interestingly, the potential for all these compounds to produce such adverse reactions generally became evident only after wide usage. These effects were not anticipated; but once they were identified the materials were either withdrawn from the commercial market or their use was restricted. As a result, the incidence of photosensitization reactions from topically applied products has decreased.

Although the incidence of such reactions has decreased, their frequency is still sufficient to warrant the screening of new chemical entities and formulations before they are introduced into commerce. In the wake of the epidemics of the 1960s, predictive test procedures evolved. Predictive testing for photosensitization potential of new products and raw materials is important for several reasons:

1. Avoid the epidemics of the past.
2. There is a rapid proliferation of new substances in our environment which are photoactive. Many are being formulated into products which will come in contact with the skin.
3. Predicting the photosensitization potential of a new chemical entity on the basis of molecular structure is not very reliable.
4. People continue to increase their exposure to sunlight. In recent years, they have even extended their exposure to ultraviolet radiation with the aid of artificial sources such as tanning beds.

5. Regulatory agencies in the United States and other countries are increasingly aware of the consequences of photosensitivity, and in many cases have formulated new guidelines for assessing the photosensitization potential of cosmetics and their components.

While in vitro and animal models exist for predicting photosensitization potential [14-20], evaluations properly performed in humans are of unique value, since they directly predict the clinical behavior of a material. This chapter will focus on human predictive patch test procedures used to assess the photosensitization potential of cosmetics.

II. OVERVIEW OF ELECTROMAGNETIC RADIATION

Electromagnetic radiation is that form of energy that can be transmitted through a vacuum and in matter. In sufficient quantity, it can be considered as transverse electromagnetic waves; i.e., rapidly alternating electric and magnetic fields which are perpendicular to each other [21]. The energy of electromagnetic radiation, in general, is described as characteristically being (1) discrete packets or particlelike, and (2) oscillating or wavelike. As early as 1900, the German physicist Max Planck developed a theory that described how tiny particles of energy, called photons, behave like waves [22].

All photons travel through space in a wavelike fashion at the constant rate of 186,282 miles per second (i.e., the speed of light). As photons of a given radiation travel, the wave formed by this radiation can be described by its frequency (number of oscillations per second) and its wavelength (physical distance between equivalent points of two consecutive waves). The relationship between frequency and wavelength of electromagnetic radiation is inverse; i.e., the shorter the wavelength, the higher the frequency.

When photons are absorbed, energy is released to the absorbing material and the photons cease to exist [21]. The amount of energy in a photon is directly proportional to the frequency of the radiation from which it is derived. Therefore, higher frequency (shorter wavelength) radiations consist of photons of higher energy than radiations of low frequency (long wavelength). This is an important principle to remember for further discussions.

Radiant energy originating from the sun is electromagnetic energy. It is a composite of wavelengths, varying from very high-energy regions (e.g., gamma and x-rays) to the low-energy regions of microwaves and radio waves. Absorption by ozone and molecule scattering in the earth's atmosphere effectively removes wavelengths less than 290 nm, whereas water vapor and CO_2 in the atmosphere absorb red and infrared radiation to further attenuate the sun's radiant energy [23]. Consequently, only a small portion of the sun's electromagnetic energy actually reaches the earth's surface. The radiation arriving at the surface contains wavelengths from about 290 to 4000 nm. This terrestrial solar spectrum is generally divided into three categories: ultraviolet (290-400 nm), visible (400-760 nm), and infrared (>760 nm). The composition of the solar spectrum at the earth's surface (sea level) is approximately 6% ultraviolet, 49% visible, and 45% infrared.

The ultraviolet region of the solar spectrum is primarily responsible for reactions in the skin. Although comprising only about 6% of the solar spectrum, ultraviolet radiation causes nearly 99% of the side effects of sunlight [24]. The ultraviolet region of the solar spectrum is further subdivided into long-wave ultraviolet (UVA: 320-400 nm) and middle-wave ultraviolet (UVB: 290-320 nm). While UVB is only about 0.5% of the total spectrum, it is high in energy and responsible for producing sunburn. On the other hand, UVA comprises nearly 10 times more of the solar spectrum, yet is 100-1000 times less effective than UVB in producing sunburn [25-27]. However, these wavelengths are of great importance because they are, in most cases, responsible for the vast majority of photosensitization reactions that occur in human skin [28].

It is generally recognized that chemical reactions are necessary for biological responses to occur. If a particular biological response is dependent upon light, then it is reasonable to assume that the chemical reactions involved in the response are a consequence of the absorption of photons by molecules in the system [23]. In fact, without absorption of radiation, photochemical reactions cannot take place. This is known as the first law of photochemistry and was developed by Grotthus and Draper in 1818.

The relative response of a biological system to different wavelengths of radiation is its action spectrum. Essentially, this is a plot of the reciprocal of the number of incident photons required to produce a given effect compared to the wavelength of the radiation employed. In developing an action spectrum, the degree and quality of the response must be the same at different wavelengths. Within the action spectrum, peaks represent wavelengths most efficient for producing the biological response (i.e., wavelengths that require the fewest photons to elicit the effect).

In photobiology, photosensitizers are molecules that absorb photons and cause a change in other molecules surrounding them [29]. The mechanism is complex, but generally involves one of the following:

Energy transfer to biological molecules such as deoxyribonucleic acid (DNA) or protein
Electron transfer
A combination of the photosensitizer with a cellular chemical

Absorption of radiation by a molecule is a relatively precise phenomenon in that specific molecular structures absorb radiation of specific wavelengths [22]. The molecular structure also determines the extent of absorption of various wavelengths or photon energies. Thus, by plotting the wavelength versus radiant energy absorbed by a given molecule, one can obtain its absorption spectrum.

Kleczkowski observed that an action spectrum in a biological system may parallel the absorption spectrum of a photosensitizer which it contains [30]. The assumption here is that the absorber or photosensitizer in the system will, on irradiation, act mostly at those wavelengths where absorption is greatest. At wavelengths where the molecule absorbs very little or not at all, the effect will be small if at all.

III. LIGHT SOURCES AND MONITORING

In predictive patch testing of cosmetics, drugs, and their ingredients for potential photosensitization reactions, the action spectrum is rarely known. If the test material is irradiated with a light source which fails to emit light in the absorption spectrum of the test material, false negatives may result. Therefore, a light source with a broad spectrum of wavelengths is essential for screening purposes. Since most known photosensitizers absorb in either the ultraviolet or visible regions, the source should contain these wavelengths.

The most logical light source, then, for predictive photopatch testing would appear to be natural sunlight. As mentioned earlier, it is a continuous spectrum abundant in ultraviolet and visible radiation. Unfortunately, natural sunlight is not a good source because the whole spectral output of terrestrial sunlight is constantly changing. Sayre and others have discussed the numerous difficulties associated with using the sun for clinical studies [31-33]. The ultraviolet portion of the solar spectrum varies considerably with season, time of day, latitude, altitude, weather patterns, and smog. Some of these same factors may prevent the conduct of the study altogether, or affect the results of an exposure. The reader should be aware that while clinical studies can be conducted in natural sunlight, they will be somewhat uncontrollable from the standpoint of light exposures.

For carefully controlled studies in which the spectral characteristics of the radiation source are known and the exposure doses can be accurately determined, artificial light sources are strongly recommended. Ideally, an artificial light source should meet several specifications:

1. Possess an emission spectrum which includes the radiation most likely to elicit the reaction under study [19]. In most cases, the emission spectrum should simulate the solar spectrum.
2. Be of sufficient intensity to achieve high radiation doses in relatively short exposure times while avoiding intense local heat and maintaining subject's comfort.
3. Have constant, uniform output.
4. Be reliable and reasonably simple to operate and maintain.
5. Safe.
6. Inexpensive.

Because no ideal light source exists, a variety of artificial sources have been developed over time. The spectral characteristics, design, advantages, and disadvantages have been reviewed in a number of publications [24,34-42]. The preferred sources for predictive photopatch testing are the gas discharge lamps (xenon arc; high-, medium-, and low-pressure mercury vapor arcs; and fluorescent tubes) because they are relatively rich in ultraviolet radiation. Arc lamps are useful mainly because of their considerable light intensity.

The primary virtue of fluorescent lamps is that they provide a fairly uniform radiant energy over a large area with very little generation of heat, and they are inexpensive. The disadvantages with these lamps are that they do not cover the whole solar spectrum and they require long irradiation times involving minutes or even hours.

In the United States, the artificial light source most often used in human predictive photopatch studies appears to be the xenon arc solar simulator originally designed by D. Berger [43]. As described, this source meets the following specifications:

It has a continuous emission spectrum in the ultraviolet and visible wavelengths which very closely approximates daylight.
When appropriately filtered, less than 1% of its total energy is contributed by wavelengths shorter than 290 nm.
No more than 5% of its erythemically effective energy is contributed by nonsolar wavelengths.
With appropriate filtering, UVB can be easily removed, yielding broad-band UVA and visible wavelengths in the source output.
The energy emitted is constant and very stable.

Figure 3.1 provides a schematic diagram of a typical xenon arc solar simulator. In this design, the light emitted from the xenon short, compact arc lamp is reflected onto a dichroic mirror mounted at a 45° angle. Dichroic mirrors act as interference filters; in this case reflecting ultraviolet and transmitting about 90% of the visible and infrared light in the form of heat. The reflected light passes through a lens that collimates the beam, thus improving the uniformity of the light intensity across the beam. In some high-powered solar simulators (1000 W or greater), a second filter, such as another dichroic mirror or a water filter, may be needed to further remove the visible and infrared light that was not removed by the first mirror. The beam of light finally travels through a cut-off filter (e.g., a 1-mm thick Schott WG320) which absorbs the short-wave radiation (UV <290 nm). A bandpass filter, such as a Schott UG-11, can also be used to absorb much of the visible and infrared light remaining in the beam at that point, yet still transmit ultraviolet and some visible wavelengths.

The result is a beam which has constant energy and is rich in UVA and UVB radiation. The filtered xenon arc lamp lacks the significant line spectral components of many of the other available lamp systems. It also has a spectral distribution which matches closely that of natural sunlight, especially in the ultraviolet region, as shown in Fig. 3.2.

For UVA exposures, the system configuration shown in Fig. 3.1 can be easily modified using either of two options. The first option simply involves the addition of a UVB absorbing filter (e.g., a 2-mm thick Schott WG345 filter) between the WG320 and UG-11 filters. This approach yields a uniform beam

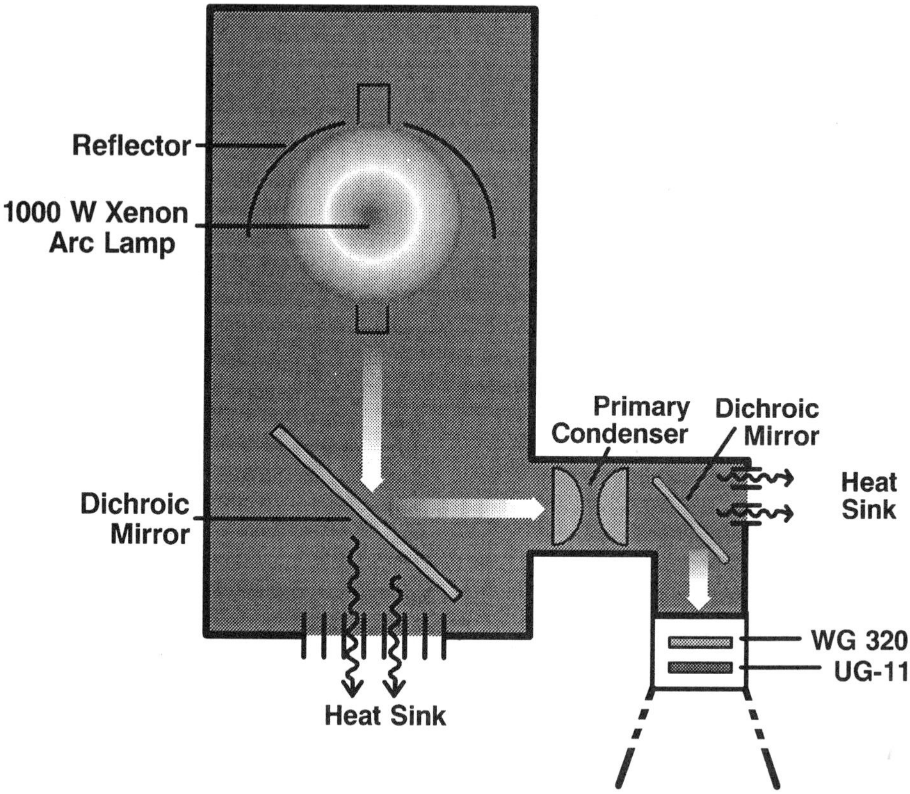

FIGURE 3.1 Configuration of a solar simulator.

of UVA light. The intensity of the UVA light is low, since it is spread over a large area (approximately 3-6 in. in diameter) The WG320 filter is not necessary in this configuration; however, wavelengths below 320 nm have been shown to cause permanent changes (called solarization) in the transmission of some filter glasses [44]. Consequently, the WG320 tends to protect the other filters from this effect. The second option also uses the system configuration shown in Fig. 3.1; however, the light beam is further directed into a 6-in. focal length focusing lens, a 2-mm thick WG345 filter, and then a UG-11 filter. The resulting beam is approximately 1 in. in diameter at a distance of about 4-5 in. from the lens. The spectral output is a very

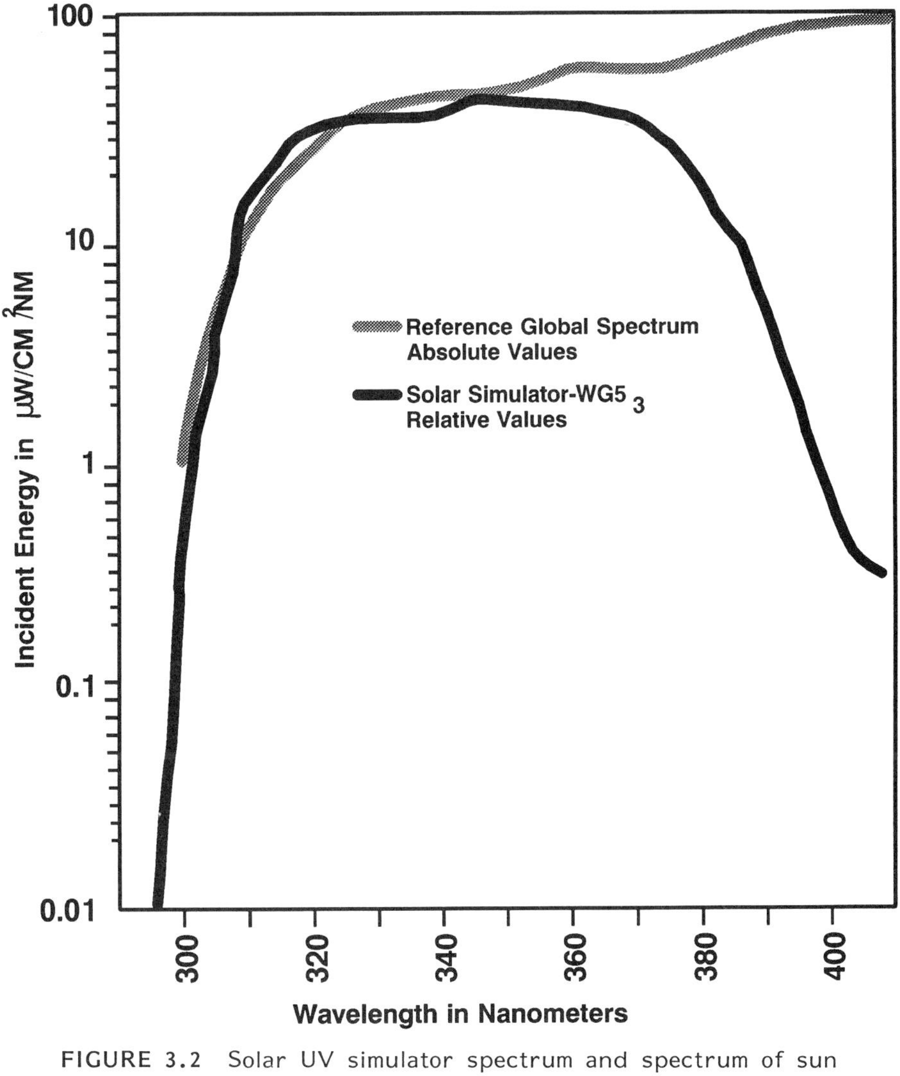

FIGURE 3.2 Solar UV simulator spectrum and spectrum of sun plus skylight. (From Ref. 43, used with permission.)

intense beam of UVA. In both options, the UG-11 filter can be removed from the system to further broaden the spectrum to include more visible radiation.

Factors contributing to the stability, spectral distribution, and life expectancy of lamps have been well documented [45-49]. Routine maintenance and monitoring of the light source used in photopatch test studies are essential for producing reliable, reproducible results. For daily monitoring, a calibrated radiometer (a device which converts absorbed radiant energy to measurable electrical energy) is generally used. The radiometer should be selected according to the emission spectrum of the light source to be used as well as the radiation intensity [24].

If the emission spectrum includes near-infrared wavelengths, the radiometer's detector should be relatively insensitive to such wavelengths. While this is not a serious problem for sources such as UVA fluorescent tubes and sun lamp fluorescent tubes (their largest output is ultraviolet radiation), solar simulators do present such a problem. Wulf and Eriksen reported that silicone photodiode detectors should generally be avoided, since their sensitivity peaks at about 900 nm [50]. They found that detectors such as vacuum photodiodes perform reasonably accurate UVA measurements when combined with simple filters.

Total irradiance of the solar simulator and uniformity of the irradiance within the beam have been measured in our laboratory on a daily basis for a number of years using a IL 700 Research Radiometer™ equipped with a calibrated cosine-corrected vacuum photodiode detector. Results obtained with this system have correlated well with those obtained using a spectroradiometer (a device used to selectively measure the radiant energy at individual wavelengths or wavebands).

Whenever major changes in the light source occur (e.g., lamp aging, lamp replacement, extended use of filters), the emission spectrum should be evaluated. As lamps age, the emission intensity of the light source generally decreases and the emission spectrum may change such that certain wavelengths are no longer emitted or are at lower intensity. Labeling or coding on lamps or packaging materials does not always accurately designate the proper lamp, especially in the case of fluorescent lamps. An excellent example of this was reported in 1976 by Forbes et al. [51]. Two blacklight fluorescent lamps having identical appearance and coding were found to have significantly different spectral distributions, and consequently different biological effectiveness. Both types of lamps also exhibited approximately equivalent intensities when measured by UVA radiometers.

Cole et al. [52] recommended that the spectral distribution of lamps be verified when they are new or being changed.

To accurately determine the emission spectrum of the source, a spectroradiometer consisting of a monochromator, radiometer, and detector is recommended. Since such equipment is quite expensive, it may not be available to the investigator. Alternatives would be to change lamps or bulbs and any filters on a routine basis so that the effects of aging and solarization are minimized, and to obtain verification of the spectral distribution from the manufacturer.

Accurate measurements of radiation intensity and spectral distribution are essential prerequisites for the evaluation of the photosensitization potential of topically applied materials such as cosmetics. Landry and Anderson [53] provide an excellent discussion of the instrumentation and sources of error associated with conducting optical radiation measurements.

IV. PRINCIPLES OF PHOTOSENSITIZATION

Photosensitivity, like skin irritation and delayed contact hypersensitivity, is a form of contact dermatitis. As noted earlier, it occurs when molecules in contact with the skin absorb photons, become excited or reactive, and subsequently initiate a series of chemical and biological reactions. Photosensitivity is a broad classification of adverse cutaneous light reactions [52]. Within this classification exist two categories of reactions: phototoxic and photoallergic. While these reactions will be briefly discussed in this section, excellent detailed discussions regarding mechanisms of action have been published by numerous authors [17,18, 20,24,28,54-59].

Clinically, it is often very difficult to distinguish between the two phenomena even though they follow completely different mechanisms. Phototoxic materials are directly toxic (i.e., injury results in cell death) to the skin, whereas photoallergic materials elicit reactions which are due to acquired immunological reactivity, presumably dependent on an antigen-antibody or cell-mediated hypersensitivity response [24,28,55,59]. Table 3.1 provides a summary comparison of some of the characteristics distinguishing phototoxic and photoallergic reactions.

A. Phototoxicity

Generally, the molecules producing phototoxicity are of low molecular weight (<500 daltons) and possess highly resonant structures which, with few exceptions, absorb UVA readily [20].

TABLE 3.1 Summary Comparison of Distinguishing Characteristics Associated with Phototoxic and Photoallergic Reactions

	Reaction	
Characteristic	Phototoxic	Photoallergic
Possibility of reaction on first exposure	Yes	No
Incubation period necessary after first exposure	No	Yes
Relative concentration of agent necessary for reaction	High	Low
Incidence in general population	Usually relatively high (theoretically 100%)	Usually very low (but some agents might reach 100%)
Flaring response at distant, previously involved sites	No	Yes
Cross reactions to structurally related agents	No	Yes, frequently
Possibility of developing persistent light reaction	No	Yes, with some agents
Cutaneous responses	Most often resembles strong sunburn	Varied morphology often consisting of erythema, edema, and possibly vesiculation

Since they are of low molecular weight, they penetrate the stratum corneum fairly well, locating predominantly in the epidermis [54]. As they absorb photons and are activated, some phototoxic agents may damage cell membranes, others may damage organelles, whereas still others may interact with DNA [57]. This damage results in cellular death and intracellular edema. If enough cell injury occurs in the skin, an inflammatory response ensues.

Phototoxic reactions are most often characterized by intense, delayed erythema (i.e., the exaggerated sunburn reaction), and are frequently followed by hyperpigmentation and peeling of the damaged skin. According to Maibach and Marzulli [18], the hyperpigmentation is very common and, clinically, is the most distinctive component of the phototoxicity reaction caused by many topically applied agents. This hyperpigmentation results from melanin formation stimulated by the acute erythema [59]. In very severe phototoxicity, edema and blistering may develop.

The acute aspects of phototoxic reactions may begin within a few minutes to hours after irradiation and tend to resolve within days to weeks. However, the hyperpigmentation may reside from many months to years. Immediate burning and/or stinging sensations (called "smarting") may also occur, as they are characteristic of some compounds (e.g., coal tar derivatives).

B. Photoallergy

Although many drugs and chemicals have been identified as photoallergens, photoallergic reactions are generally uncommon [13]. Many of these compounds are structurally similar to agents that cause allergic contact dermatitis or elicit phototoxic reactions. Photoallergens, like contact allergens, are generally lipid-soluble, low molecular weight substances [9]. As with phototoxic compounds, photoallergens possess highly resonant structures which readily absorb over a wide spectrum consisting predominantly of UVA wavelengths. Less energy is generally required to elicit a photoallergic reaction as compared to a phototoxic reaction [59].

In photoallergy, the light energy does not excite the skin. Instead, ultraviolet or visible radiation photochemically converts an immunologically inactive molecule (the photosensitizer) into its active form [9,20,55]. The manner by which this occurs is not clear. Many mechanisms have been proposed, but only two seem to be widely supported [60,61].

One theory suggests that radiation is absorbed by the photosensitizer which subsequently is photochemically converted into one or more stable reaction products. It is one of these stable metabolites that then serves as the hapten. This hapten then combines with proteins in the skin to form the complete antigen. The drug sulfanilamide is believed to follow this mechanism [9].

According to the second theory, within the skin, radiant energy elevates the photosensitizer to an unstable, excited state. Interaction between the short-lived, excited molecules and adjacent cutaneous proteins results in protein binding to form the complete antigen. This mechanism differs from the first in that no stable metabolite is formed prior to protein binding. Para-aminobenzoic acid and several of the halogenated salicylanilides have been shown to produce excited-state molecules upon exposure to UVA radiation [9].

Independent of the pathway, once the complete antigen has been formed, the reaction proceeds in the same manner as other delayed hypersensitivity reactions. The antigen is processed by macrophages, and presented to T lymphocytes. Proliferation of sensitized T lymphocytes then occurs in the regional lymph nodes. Upon subsequent environmental exposure to the photoantigen, the sensitized T cells recognize it as foreign. They release soluble mediators that increase vascular permeability, recruit additional inflammatory cells, and ultimately produce acute dermatitis [20,55]. Like allergic contact dermatitis, the immune system's sensitivity to the photoantigen is long lived.

Clinically, photoallergic contact reactions are identical to any other type of allergic contact dermatitis [28]. Reactions may range from erythema to papules or even blisters. Edema resulting from accumulation of cellular infiltrate in the dermis may also accompany these reactions. Eruptions often extend beyond the exposure site. Intense itching may also be associated with photoallergic reactions.

Whereas the acute aspects of phototoxic reactions typically present soon after irradiation, photoallergic reactions seldom occur after the first exposure to the photoallergen. Since photoallergy is immulological in nature, multiple exposures to the photoactivated allergen and a latent period of several days to weeks are typically required before the immune system is ready to respond. Once an individual has been sensitized, subsequent exposures to the photoallergen and light will elicit cutaneous responses quite rapidly—often within 24-72 hr. Occasionally, the response is not manifested for several days. Photoallergic reactions tend to peak within 48-72 hr after they begin and subside over the next week or so if exposure to the photoallergen and light is discontinued.

V. PROTOCOL DESIGN AND RATIONALE

Many testing procedures have been devised to predict the phototoxicity and photoallergenicity potential of cosmetics and drugs [6,10,14,18,57,62-75]. Most of these techniques have lacked specificity and sensitivity. In fact, no single system yet described has been completely successful in predicting the clinical photosensitizing potential of all chemicals in man.

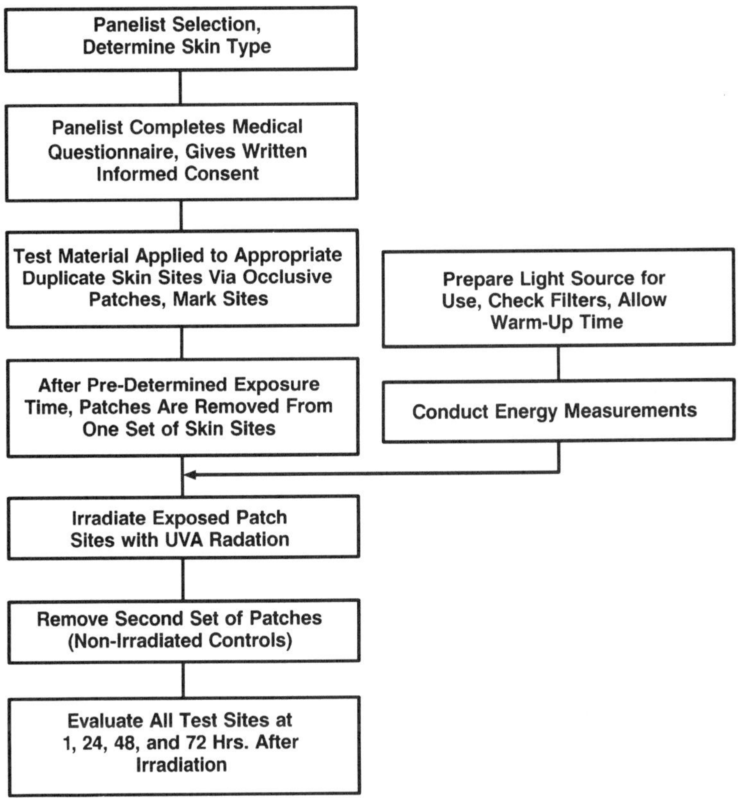

FIGURE 3.3 Phototoxicity predictive patch procedure.

In general, the methods currently in use have evolved from predictive patch techniques used to evaluate skin irritation and contact allergy potential [76,77]. Although none of the procedures has het been standardized, each of the methods generally follows a similar pattern. The general patterns are outlined in the flow charts in Figs. 3.3 and 3.4 for phototoxicity and photoallergy, respectively.

A. Phototoxicity

The following is a description of the phototoxicity patch test procedure routinely and successfully used by our laboratory.

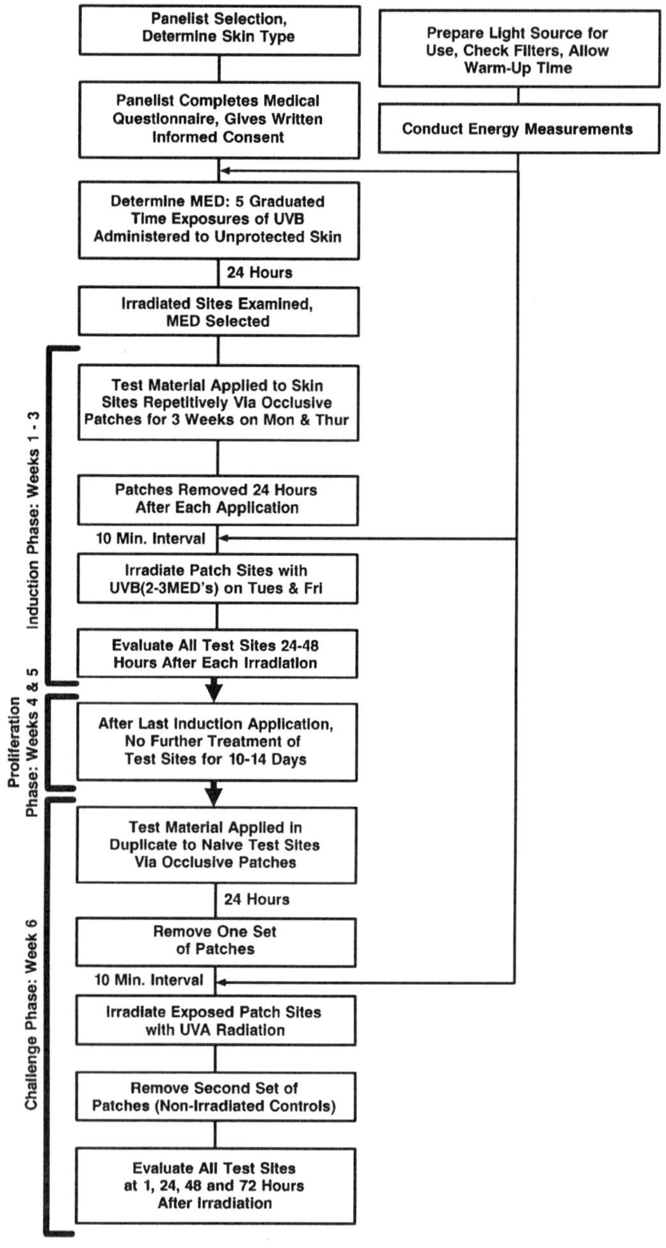

FIGURE 3.4 Photoallergy predictive patch procedure.

TABLE 3.2 Skin Types

Skin type	Sunburn and tanning history[a]	Example
I	Always burns easily; never tans	People most often with fair skin, blue eyes, freckles; unexposed skin is white[b]
II	Always burns easily; tans minimally	People most often with fair skin, red or blond hair, blue, hazel, or even brown eyes; unexposed skin is white
III	Burns moderately; tans gradually and uniformly (light brown)	Normal average white person; unexposed skin is white
IV	Burns minimally; always tans well (moderate brown)	People with white or light brown skin, dark brown hair, dark eyes (e.g., Mediterraneans, Mongoloids, Orientals, Hispanics); unexposed skin is white or light brown
V	Rarely burns, tans profusely (dark brown)	Brown-skinned persons (e.g., Amerindians, East Indians, Hispanics); unexposed skin is brown
VI	Never burns; deeply pigmented (black)	Blacks (e.g., African and American Negros, Australian and South Indian aborigines); unexposed skin is black

[a]Based on first 30-60 min sun exposure after winter season or no sun exposure.

[b]These individuals may be of Celtic descent (Irish or Scottish); others may have dark hair or brown eyes.

In general, this procedure follows the approach used by those referenced above, especially that of Kaidbey [10,14,73].

Initiation of the predictive patch test begins with enrollment of the volunteer human panel. The panel size ranges from 10 to 25 male/female subjects. The panelists are recruited from a normal, healthy population between the ages of 18 and 50 and are screened for their skin sensitivity to sun-induced damage and their medical background.

The susceptibility of the skin to sun-induced damage is classified into six skin types (Table 3.2). These skin types are based upon the individual's tendency to sunburn and subsequently tan. In most cases, volunteers with skin type I, II, or III are best suited to participate in such studies.

Individuals with a history of abnormal responses to sunlight as well as those with skin diseases such as eczema, psoriasis, or skin cancer are excluded from participating in the study. Anyone taking medications which are known to produce abnormal responses to sunlight are also excluded. Panelists are restricted from using anti-inflammatory agents during the course of the study, as these could influence the skin's responsiveness. Exposure to extraneous light by way of sun bathing or tanning beds is also prohibited.

Once the panelists have been selected, each person is informed about the nature of the procedure and any potential hazards associated with participating in the study. As in any clinical evaluation, written informed consent must be obtained from each subject.

After qualifying for the study, each subject is patched with the test material. Typically, the untanned midportion of the back is used for testing; however, the volar surface of the forearms can also be used if they are free of sunburn, suntan, scars, or other conditions which might interfere with the evaluation of the skin responses.

For evaluation of phototoxicity, a single application of test material is made in duplicate to each subject. One set will be irradiated while the other set will not be irradiated. Approximately 0.3 ml of sample is applied to a nonwoven cotton pad (Webril) which has an area of about 4 cm^2. The concentration of the test material is usually the highest concentration which is nonirritating to the skin. The treated patches are held in place on the skin with impermeable adhesive tape (e.g., Blenderm) so that an occlusive environment is achieved. Control patches are also included. These are prepared with the test material vehicle. If vehicle is not available, distilled water or physiological saline is recommended.

After approximately 24 hr, one set of patches is removed from the skin test sites. Within 10 min, the sites are irradiated with 16-20 J/cm^2 of UVA. This dose is generally less than that which will produce erythema on nontanned white skin. The source used in this laboratory is the 1000-W xenon arc solar simulator filtered as described earlier. While irradiating these sites, care must be taken to protect the nonirradiated sites from inadvertent exposure to UVA.

At the completion of the light exposures, the remaining duplicate patch sites are uncovered. All sites are then examined for skin responses at approximately 1, 24, 48, and 72 hr after patch removal. Unsolicited, subjective comments offered by the panelists are also recorded at these intervals.

B. Photoallergy

The procedure used by our laboratory to evaluate the photoallergy potential of cosmetics and topical drugs is a modification of the photomaximization procedure developed by Kaidbey and Kligman [74]. This procedure is essentially a 6-week test consisting of three phases: induction (weeks 1-3), rest or proliferation (weeks 4 and 5), and challenge (week 6). These phases correlate with those generally associated with the sensitization process.

Enrollment of 25 panelists for this predictive patch test is essentially the same as described previously for the phototoxicity procedure. Since this photoallergy procedure incorporates exposures of UVB, an additional step in the enrollment process is the determination of each subject's minimal erythema dose (MED). The MED is determined by administering a geometric series of five UVB exposures in which each dose is 25% greater than the previous one. The exposures are approximately 1 cm^2 each and are conducted on the untanned midback. About 24 hr later, these sites are examined for the dose which produced minimal but uniform erythema with a distinct border.

During the induction phase, a series of six applications of the test material is made repetitively to the same site on the untanned midback of each subject. These applications are usually performed on Monday and Thursday of each of three consecutive weeks. Each application consists of approximately 0.3 ml of the sample on a nonwoven cotton pad (Webril) which has an area of about 4 cm^2. The concentration of the test material is usually the highest concentration which is nonirritating and nonphototoxic to the skin. The treated patches are held in place on the skin with impermeable adhesive tape (e.g., Blenderm) so that an occlusive environment is achieved. In complex formulations, a vehicle control site should also be incorporated as a test site.

Twenty-four hours after each induction application (each Tuesday and Friday) the patches are removed. Within 10 min, a 1 cm^2 area of each skin test site is exposed to twice the MED of UVB. The source, a filtered 1000-W xenon arc solar simulator,

TABLE 3.3 Scoring Scale for Photoallergy Predictive Patch Test

Score	Description
0	No evidence of any reaction
1	Reaction readily visible but mild unless letter grade appended (see grades E and F below). Mild reactions include weak but definite erythema, and weak superficial skin responses such as glazing, cracking, or peeling
2	Definite papular response (append E or F or S if appropriate)
3	Definite edema (append E or F or S if appropriate)
4	Definite edema and papules (append E or F or S if appropriate)
5	Vesicular/bullous eruption
E	If the letter E is appended to the numerical grade, it documents the presence of strong erythema at the patch site
F	If the letter F is appended to the numerical grade, it documents the presence of strong effects on superficial layers of the skin, which include fissures, a film of dried serous exudate, small petechial erosions, and/or scabs
S	If the letter S is appended to the numerical grade, it documents the presence of a reaction spreading beyond test site

was described earlier. Between irradiation and the next application, sites are left uncovered.

Following irradiations 1, 3, and 5 skin responses are evaluated approximately 24 hr after completion of irradiation. Skin responses to irradiations 2, 4, and 6 are examined approximately 72 hr after irradiation.

As in predictive patch procedures for delayed contact hypersensitivity, the photoallergy procedure incorporates a 10- to 14-day rest period. During this period, no applications of the test material are performed nor is there any intentional ultraviolet exposure. Panelists are required to adhere to all restrictions covered during enrollment.

At challenge, a single application of duplicate contact patches of the test material and vehicle control is made to naive sites. This application is conducted in the same manner as those of the induction phase. About 24 hr later, one set of patches is removed, and within 10 min, these sites are irradiated with 16-20 J/cm^2 of UVA. The light source is the same as that used for the phototoxicity test. During irradiation, care is taken to avoid irradiating the remaining covered patch sites, as those are the nonirradiated controls.

When irradiation is completed, the remaining sites are uncovered. All sites are then examined for skin responses at approximately 1, 24, 48, and 72 hr after patch removal. Skin responses are evaluated according to the scoring scale shown in Table 3.3. Unsolicited, subjective comments offered by the panelists are also recorded at these intervals.

VI. SOME FACTORS FOR CONSIDERATION BEFORE STUDY EXECUTION

Adequate evaluation of photosensitization potential depends on a thorough understanding of the factors which influence the response. Unfortunately, experience with predictive photopatch testing in humans has been limited and factors that influence the final response, such as test material concentration, vehicle, and ultraviolet dose, need to be further investigated. Discussed below are several factors which should be considered during the planning stages of a predictive photopatch test.

A. Sample Concentrations

As indicated earlier, phototoxic agents typically require relatively high concentrations to cause skin reactions. If the test concentration is too low, false negative patch results may be obtained. Some information on phototoxic concentration from animal models is often helpful in determining what concentration to use in human tests.

In the case of predictive patch testing for photoallergy, as with contact allergy, induction and elicitation are dose related. Hence, the higher the test concentration, the greater the probability of identifying the photoallergy potential. Care must be taken, however, that the concentration selected is not inherently irritating or even phototoxic. Excessive concentrations may make the distinction between photoallergy and irritation or phototoxicity difficult at challenge. Results of human or animal skin irritation or phototoxicity screens should provide assistance in selecting suitable levels.

B. Vehicles

Vehicles used to solubilize a test material are very important because they can potentially alter percutaneous penetration. Test agent release and penetration can be either increased or decreased depending on the properties of both the agent and the vehicle. Kaidbey and Kligman found that the photosensitizing capacity of crude coal tar, methoxsalen, and chlorpromazine is strongly influenced by the choice of vehicle [78]. In general, emulsion-type creams were superior to polyethylene glycol and petrolatum in enhancing penetration.

When dilution is necessary, the vehicle itself should not cause skin dermatitis. Vehicles which are volatile (e.g., ethanol, acetone) tend to be irritating and in some cases allergenic. If time is permitted for the vehicle to evaporate sufficiently from the patches, the potential for dermatitis is usually diminished.

C. Patch Contact Time

While many chemicals possess a considerable lag time before significant amounts of material are percutaneously absorbed, Maibach and Marzulli [18] indicate that this is not always the case. They report that for the perfume ingredient bergamot oil the optimum light exposure interval in humans is 1-2 hr after application to the skin. If light exposure is conducted 24 hr after application, no response occurs.

Patch contact times for phototoxicity evaluations vary widely in the literature. Kaidbey and Kligman suggest 6-hr contact time in their procedure [73]. In our laboratory, 24-hr contact time has been routinely used with considerable success. This may be related to the fact that by saturating the patch pad with 0.3 ml of test material, the skin site is constantly exposed to a fresh supply of the test material over the 24-hr contact period. In any case, the time factor should be taken into account when planning a predictive assay.

D. Light Exposure

While the majority of phototoxic and photoallergenic agents are activated by UVA radiation, it is important to remember that not all agents are activated by these wavelengths. In fact, some synthetic dyes, such as eosin and rose bengal, require visible light for activation [73]. Consequently, the investigator should attempt to use a light source of relatively broad spectrum to avoid missing the activation wavelengths of the test material, especially new materials with unknown action spectra.

Although 2-3 J/cm^2 of UVA is very likely the photoactivating dose for most phototoxic chemicals, Kligman and Kaidbey suggest that the suberythemic dose of 20 J/cm^2 is appropriate for phototoxicity evaluations [57]. In fact, this dose is in line with the environmental UVA exposure an individual might experience in the summer. On a cloudless day with the sun directly overhead, the solar UVA irradiance is about 5 mW/cm^2 [35]. At this rate, a person could be exposed to between 15 and 48 J/cm^2 of UVA over the course of a midsummer day [10,79].

Regarding the dose of UVA required to elicit photoallergy responses, exceedingly small doses of UVA appear to be sufficient to provoke such responses in strongly sensitized individuals [80,81]. In the photomaximization procedure, Kaidbey and Kligman [74] report the use of 4 J/cm^2 of UVA at challenge. While this dose has been used to successfully identify a wide variety of photoallergens, our laboratory routinely uses 16-20 J/cm^2 unless animal predictive tests or other information indicate otherwise for a given test material. This dose is very likely excessive; however, since the predictive patch test is used to evaluate a wide variety of compounds, some of which may be very weak photosensitizers, such a dose simply provides an additional level of protection and assurance that there will be sufficient UVA to detect even weak photoallergens.

E. Scoring Skin Reactions

A very important aspect of study execution involves scoring the skin reactions. Individuals responsible for scoring should be adequately trained and have a thorough understanding of the principles of phototoxicity and photoallergy. This can be of great help in interpreting the data. Remember, the data are only as reliable as those who have generated them.

Phototoxic responses vary in quality and in time pattern with different types of agents. Material such as coal tar and some dyes produce an immediate wheal and flare reaction, accompanied by a burning and stinging sensation during or shortly after irradiation. Consequently, the investigator should watch carefully for such reactions as well as the classic delayed "sunburn" reaction and be aware that they are also characteristic of phototoxicity. The investigator should also inspect the test sites at several intervals so as not to overlook a positive response.

In the conduct of the induction phase of the photoallergy patch test, intense inflammatory reactions will likely develop by the second induction exposure. This is a result of the sunburn produced by the repeated 2 × MED UVB exposures. Erythema, edema, and crusting are often present by the end of the second week, while scaling and tanning usually develop by the end of the third induction week. The investigator should be aware that these are normal reactions for such repeated UVB exposures and are usually well tolerated by the panelist.

VII. INTERPRETATION

The interpretation of predictive patch test results is a subjective process. The investigator examines the response patterns and the intensity of the responses, compares these with the nonirradiated and vehicle controls, and then adds to this any other information derived from the panelist (e.g., subjective comments of burning, itching, stinging) or previous experience with similar materials. With this information, the investigator then begins to put the puzzle together to draw conclusions.

Interpretation of photopatch test results is in most cases very straightforward. In general, test materials causing dermatitis at irradiated sites but not at nonirradiated sites are considered positive. Difficulties arise, however, when owing to either ordinary skin irritation, contact allergy, or some other reason, the unirradiated controls exhibit a skin response which is similar to or weaker than that of the irradiated sites. The following discussion is intended to serve as a general guide to the interpretation of phototoxicity and photoallergy patch test data.

In the predictive patch test for phototoxicity, an immediate wheal and flare reaction or a delayed erythema reaction only at the irradiated test material site constitutes a definite phototoxic response. As described earlier, the erythema may be accompanied by other skin reactions. If, however, the nonirradiated site exhibits a similar response, then the test material is probably not phototoxic but irritant in nature.

Photocontact sensitization as a rule is characterized by erythema and edema only at the irradiated sites during challenge. Stronger skin reactions may also occur during the challenge. Test materials exhibiting nearly equivalent responses at both irradiated and nonirradiated sites are most often contact allergens rather than photoallergens. If the irradiated site exhibits a distinctly more intense response than the nonirradiated site, it is possible that a dual sensitivity exists; i.e., both contact and photocontact allergy. In such cases, photopatch testing with serial dilutions of the sensitizer is recommended [81]. A positive result at the irradiated site of a concentration that fails to elicit a response at the respective unirradiated test site is suggestive of dual sensitization.

A phototoxic reaction at challenge may also occur and be misinterpreted as a photoallergic response. For this reason, test materials should be screened beforehand for phototoxic potential so that the concentration to be used in the photoallergy test can be selected below the threshold dose for phototoxicity.

It is important to realize that the results of such tests, especially those of the photomaximization test, do not indicate the incidence of photosensitization which is likely to occur in the general population under normal usage conditions. These tests simply identify the *potential* for such reactions to develop with a given test material. Furthermore, negative results do not necessarily mean that the potential is nonexistent, since very weak or marginally active agents may not be detected. Instead, no reactions in a panel of 25 subjects is reasonable assurance that the potential for photoallergy to be induced by a given test material is relatively low.

SUMMARY

Most of the known photosensitizing agents were identified only after they had reached widespread use. Each year, as a result of consumer demand, increasing numbers of cosmetic products, drugs, household products, and personal care products are developed and produced. New chemicals are continuously introduced into the market, many of these, such as sunscreens, dyes, fragrances, and antimicrobials, often have molecular structures capable of absorbing ultraviolet radiation. The possibility that consumers may develop phototoxic or photoallergic skin reactions to these materials should be an important consideration in the development and premarket safety screening of new chemicals and finished products.

Predictive patch tests in humans are, like the animal tests, an important aspect of the safety evaluation of these new materials. While this discussion has focused on human predictive patch tests for photosensitization potential, in no way is it implied that animal tests for such responses should be abandoned. Certainly there is an advantage in evaluating products in humans because the uncertainty of extrapolation from the animal model is avoided. However, testing in animal models is still essential, especially as a preliminary step to testing in humans. Results from animal tests for phototoxicity and photoallergy can be quite valuable to the investigator in both study design and protection of human subjects from undue risk.

Our goal for the future should be to develop a better understanding of the factors which influence predictive tests for phototoxicity and photoallergy and to use this new knowledge to design more reliable methods in both animals and humans.

REFERENCES

1. Peterkin, G. G., Sulphonamide rashes. An analysis of 500 cases in North Africa and Italy, Br. Med. J., 2:1-6, 1945.
2. Wilkinson, D. S., Photodermatitis due to tetrachlorosalicylanilide, Br. J. Dermatol., 73:213-219, 1961.
3. Calnan, C. D., Harman, R. R. M., and Wells, G. C., Photodermatitis from soaps, Br. Med. J., 2:1266, 1961.
4. Epstein, J. H., Photoallergy and photoimmunology. In Dermatologic Immunology and Allergy (J. Stone, ed.), Mosby, St. Louis, 1985.
5. Smith, S. Z., and Epstein, J. H., Photocontact dermatitis to halogenated salicylanilides and related compounds, Arch. Dermatol., 113:1372-1374, 1977.
6. Kaidbey, K. H., and Kligman, A. M., Photocontact allergy to 6-methylcoumarin, Contact Dermatitis, 4:277-282, 1978.
7. Larsen, W. G., Perfume dermatitis, Am. Acad. Dermatol., 12:1, 1978.
8. Giovinazzo, V. J., Harber, L. C., Bickers, D. R., and Armstrong, R. B., Photoallergic contact dermatitis to musk ambrette, Arch. Dermatol., 117:344-348, 1981.
9. Elmets, C. A., Drug-induced photoallergy. In Dermatologic Clinics: Photosensitivity Diseases (V. A. DeLeo, ed.), Saunders, Philadelphia, 231-241, 1986.
10. Kaidbey, K. H., and Kligman, A. M., Phototoxicity to a sunscreen ingredient, Arch. Dermatol., 114-547-549, 1978.
11. Mathias, C. G. T., Maibach, H. I., and Epstein, J., Allergic contact photodermatitis to para-aminobenzoic acid, Arch. Dermatol., 114:1665-1666, 1978.

12. Holzle, E., and Plewig, G., Photoallergische kontaktdermatitis durch benzophenanhaltige sonnenschutzpraparate, Hautarzt, 33:391-393, 1982.

13. Nater, J. P., and deGroot, A. C., Phototoxic and photoallergic contact dermatitis. In Unwanted Effects of Cosmetics and Drugs Used in Dermatology, Excerpta Medica, Amsterdam, 90-98, 1983.

14. Kaidbey, K., Assessment of topical photosensitizers in humans. In Safety and Efficacy of Topical Drugs and Cosmetics (A. M. Kligman and J. J. Leyden, eds.), Grune & Stratton, New York, 213-219, 1982.

15. Harber, L. C., Armstrong, R. B., and Ichikawa, H., Current status of predictive animal models for drug photoallergy and their correlation with drug photoallergy in humans, J. Natl. Cancer Inst., 69:237-244, 1982.

16. Maurer, T., Predictive animal models for photoallergenicity testing. In Contact and Photocontact Allergens, Marcel Dekker, New York, 141-156, 1983.

17. Kornhauser, A., Wamer, W., and Giles, A., Jr., Light-induced dermal toxicity: Effects on the cellular and molecular level. In Dermatotoxicology (F. M. Marzulli and H. I. Maibach, eds.), Hemisphere Publishing, Washington, D.C., 323-355, 1983.

18. Maibach, H. I., and Marzulli, F. N., Phototoxicity (photoirritation) of topical and systemic agents. In Dermatotoxicology (F. M. Marzulli and H. I. Maibach, eds.), Hemisphere Publishing, Washington, D.C., 375-389, 1983.

19. Gabriel, K. L., Forbes, P. D., and Davies, R. E., Phototoxicity. In Photobiology of the Skin and Eye (E. M. Jackson, ed.), Marcel Dekker, New York, 33-39, 1986.

20. Stephens, R. J., and Bergstresser, P. R., Fundamental concepts in photoimmunology and photoallergy. In Photobiology of the Skin and Eye (E. M. Jackson, ed.), Marcel Dekker, New York, 41-66, 1986.

21. Parrish, J. A., Photobiology and immunology. In The Effect of Ultraviolet Radiation on the Immune System (J. A. Parrish, ed.), Johnson & Johnson Baby Products Company, 3-20, 1983.

22. Parrish, J. A., Anderson, R. R., Urbach, F., and Pitts, D., The spectrum of electromagnetic radiation: UVA in perspective. In UV-A Biological Effects of Ultraviolet Radiation with Emphasis on Human Responses to Longwave Ultraviolet, Plenum Press, New York, 1-6, 1978.

23. Megaw, J. M., and Drake, L. A., Photobiology: An overview. In Photobiology of the Skin and Eye (E. M. Jackson, ed.), Marcel Dekker, New York, 1-31, 1986.

24. Maurer, T., Photocontact allergenicity. In Contact and Photocontact Allergens, Marcel Dekker, New York, 111-123, 1983.

25. Parrish, J. A., Anderson, R. R., Urbach, F., and Pitts, D., Immediate and short-term biologic effects of ultraviolet radiation on normal skin. In UV-A Biological Effects of Ultraviolet Radiation with Emphasis on Human Responses to Longwave Ultraviolet, Plenum Press, New York, 107-139, 1978.

26. Whitman, G. B., Leach, E. E., DeLeo, V. A., and Haber, L. C., Comparative study of erythema response to UVA radiation in humans and guinea pigs. In The Biological Effects of UVA Radiation (F. Urbach and R. W. Gange, eds.), Praeger, New York, 79-86, 1986.

27. Cavallo, J., and DeLeo, V. A., Sunburn. In Dermatologic Clinics: Photosensitivity Diseases (V. A. DeLeo, ed.), Saunders, Philadelphia, 181-187, 1986.

28. Epstein, J. H., Photocontact allergy in humans. In Dermatotoxicology (F. M. Marzulli and H. I. Maibach, eds.), Hemisphere Publishing, Washington, D.C., 391-404, 1983.

29. Kochevar, I. E., Photobiology-basic science. In Dermatologic Clinics: Photosensitivity Diseases (V. A. DeLeo, ed.), Saunders, Philadelphia, 171-179, 1986.

30. Kleczkowski, A., Action spectra and their interpretation. In Research Progress in Organic Biological and Medicinal Chemistry (U. Gallo and L. Santamaria, eds.), North Holland, Amsterdam, 1972.

31. Sayre, R. M., Desrochers, D. L., Marlowe, E., and Urbach, F., Correlation of indoor solar simulator with natural sunlight, Arch. Dermatol., 114:1649-1651, 1978.

32. Morton, J. J. P., and Murphy, E. G., Sun protection factors: Regulatory issues and methods of determination. In Photobiology of the Skin and Eye (E. M. Jackson, ed.), Marcel Dekker, New York, 85-103, 1986.

33. Billhimer, W. L., Human sunscreen evaluation: Protection from sunburn, Cosmet. Toilet., 102:83-89, 1987.

34. Harber, L. C., Bickers, D. R., Epstein, J. H., Pathak, M. A., and Urbach, F., Light sources used in photopatch testing. In Sunlight and Man (T. B. Fitzpatrick, ed.), University of Tokyo Press, Tokyo, Japan, 1974.

35. Parrish, J. A., Anderson, R. R., Urbach, F., and Pitts, D., Sources of UV-A. In UV-A Biological Effects of

Ultraviolet Radiation with Emphasis on Human Responses to Longwave Ultraviolet, Plenum Press, New York, 1978, pp. 15-25.

36. Johnson, B. E., and Mackenzie, L. A., Techniques used in the study of the photodermatoses. In Seminars in Dermatology: Photodermatoses (W. Frain-Bell, A. J. Rook, and H. I. Maibach, eds.), Thieme-Stratton, New York, 1982, pp. 218-225.

37. Jagger, J., Appendix A. Instrumentation. In Solar-UV Actions on Living Cells, Praeger, New York, 1985, pp. 159-163.

38. Magnus, I. A., Methods in photobiological research. In Methods in Skin Research (D. Skerrow and C. J. Skerrow, eds.), Wiley, New York, 475-498, 1985.

39. Mutzhas, M. F., UVA-emitting light sources. In The Biological Effects of UVA Radiation (F. Urbach and R. W. Gange, eds.), Praeger, New York, 10-29, 1986.

40. Diffey, B. L., The spectral emissions from ultraviolet radiation lamps used in dermatology, Photodermatology, 3: 179-185, 1986.

41. Anderson, R. F., Artificial light sources. In Dermatologic Clinics: Photosensitivity Diseases (V. A. DeLeo, ed.), Saunders, Philadelphia, 203-215, 1986.

42. Hader, D., and Tevini, M., Light sources. In General Photobiology, Pergamon Press, Oxford, England, 1987, pp. 26-30.

43. Berger, D., Specification and design of solar ultraviolet simulators, J. Invest. Dermatol., 53(3):192-199, 1969.

44. Optical Glass Filters, Schott Glass Technologies, Inc., catalog, Duryea, Pennsylvania, p. 29.

45. Henderson, S. T., and Marsden, A. M., Lamps and Lighting, Edward Arnold, London, 1972, pp. 213-269.

46. Mullen, P. A., Kinmouth, R. A., and Searle, N. Z., Spectral energy distributions and aging characteristics of fluorescent sunlamps and blacklights, J. Testing Eval., 3:15-20, 1975.

47. Philips, R., Sources and Applications of Ultraviolet Radiation, Academic Press, London, 1983, pp. 206-263.

48. Hartmann, K. M., Action spectroscopy. In Biophysics (W. Hoppe, W. Lohmann, H. Markl, and H. Ziegler, eds.), Springer-Verlag, Berlin, 1983.

49. Diffey, B. L., The stability of light sources: Implications for photobiological studies, Photochem. Photobiol., 47:317-320, 1988.

50. Wulf, H. C., and Eriksen, P., Choosing the proper filter/detector combination for the measurement of UVA and UVB radiation from various light sources, Photodermatology, 2:264-271, 1985.

51. Forbes, P., Davies, R., D'Aloisio, L., and Cole, C., Emission spectra differences in fluorescent blacklight lamps, Photochem. Photobiol., 24:613-615, 1976.

52. Cole, C. A., Forbes, P. D., and Davies, R. E., Different biologic effectiveness of blacklight fluorescent lamps available for therapy with psoralens plus ultraviolet A, J. Am. Acad. Dermatol., 11:599-606, 1984.

53. Landry, R., and Anderson, F. A., Optical radiation measurements: Instrumentation and sources of error, J. Natl. Cancer Inst., 69:155-161, 1982.

54. Parrish, J. A., Anderson, R. R., Urbach, F., and Pitts, D., Adverse cutaneous reactions to UV-A. In UV-A Biological Effects of Ultraviolet Radiation with Emphasis on Human Responses to Longwave Ultraviolet, Plenum Press, New York, 1978, pp. 15-25.

55. Harber, L. C., Bickers, D. R., Armstrong, R. B., and Kochevar, I. E., Drug photosensitivity: Phototoxic and photoallergic mechanisms. In Seminars in Dermatology: Photodermatoses (W. Frain-Bell, A. J. Rook, and H. I. Maibach, eds.), Thieme-Stratton, New York, 1982, pp. 183-195.

56. Epstein, J. H., Chemical phototoxicity in humans, J. Natl. Cancer Inst., 69:265-268, 1982.

57. Kligman, A. M., and Kaidbey, K. H., Human models for identification of photosensitizing chemicals, J. Natl. Cancer Inst., 69:269-272, 1982.

58. Harber, L. C., Shalita, A. R., and Armstrong, R. B., Immunologically mediated contact photosensitivity in guinea pigs. In Dermatotoxicology (F. M. Marzulli and H. I. Maibach, eds.), Hemisphere Publishing, Washington, D.C., 357-373, 1983.

59. Epstein, J. H., Phototoxicity and photoallergy in man, J. Am. Acad. Dermatol., 8:141-147, 1983.

60. Storck, H., Photoallergy and photosensitivity, Arch. Dermatol., 91:469-482, 1965.

61. Harber, L. C., and Baer, R. L., Pathogenic mechanisms of drug-induced photosensitivity, J. Invest. Dermatol., 58:327-341, 1972.

62. Curwen, W. L., and Jillson, O. F., Light hypersensitivity, J. Invest. Dermatol., 34:207-212, 1960.

63. Burdick, K., Phototoxicity of Shalimar perfume, Arch. Dermatol., 93:424-425, 1966.

64. Peck, S. M., and Vinson, L. J., Use of two prophetic patch tests for the practical determination of photosensitization potential of widely used deodorant soaps, J. Soc. Cosmet. Chem., 18:361-365, 1967.

65. Kligman, A., and Breit, R., The identification of phototoxic drugs by human assay, J. Invest. Dermatol., 51:90-99, 1968.

66. Willis, I., and Kligman, A. M., The mechanism of photoallergic contact dermatitis, J. Invest. Dermatol., 51:378-384, 1968.

67. Giovacchini, R. P., Premarket testing procedures of a cosmetic manufacturer, Toxicol. Appl. Pharmacol., 3:13-18, 1969.

68. Marzulli, F. N., and Maibach, H. I., Perfume phototoxicity, J. Soc. Cosmet. Chem., 21:685-715, 1970.

69. Frost, P., Weinstein, G. D., and Gomez, E. C., Methacycline and demechlocycline in relation to sunlight, JAMA, 216:326-329, 1971.

70. Harber, L. C., Baer, R. L., and Bickers, D. R., Technique of evaluation of phototoxicity and photoallergy in biologic systems, including man, with particular emphasis on immunologic aspects. In Sunlight and Man (T. B. Fitzpatrick, ed.), University of Tokyo Press, Tokyo, Japan, 1974.

71. Forbes, P. D., Urbach, F., and Davis, R. E., Phototoxicity testing of fragrance raw materials, Food Cosmet. Toxicol., 15:55-60, 1977.

72. Kaidbey, K. H., and Kligman, A. M., Identification of systemic phototoxic drugs by human intradermal assay, J. Invest. Dermatol., 70:272-274, 1978.

73. Kaidbey, K. H., and Kligman, A. M., Identification of topical photosensitizing agents in humans, J. Invest. Dermatol., 70:149-151, 1978.

74. Kaidbey, K. H., and Kligman, A. M., Photomaximization test for identifying photoallergic contact sensitizers, Contact Dermatitis, 6:161-169, 1980.

75. Kaidbey, K. H., and Kligman, A. M., Identification of contact photosensitizers by human assay. In Current Concepts in Cutaneous Toxicity (V. A. Drill and P. Lazar, eds.), Academic Press, New York, 55-68, 1980.

76. Shelanski, H. A., and Shelanski, M. V., A new technique of human patch tests, Proc. Sci. Sect. Toilet Goods Assoc., 19:46, 1953.

77. Kligman, A. M., The identification of contact allergens by human assay. III. The maximization test. A procedure for screening and rating contact sensitizers, J. Invest. Dermatol., 47:393-409, 1966.

78. Kaidbey, K. H., and Kligman, A. M., Topical photosensitizers—Influence of vehicles on penetration, Arch. Dermatol., 110:868-870, 1974.

79. Maurer, T., Experimental contact photoallergenicity: Guinea pig models, Photodermatology, 1:221-231, 1984.

80. Willis, I., and Kligman, A. M., Photocontact allergic reactions: Elicitation by low doses of long ultraviolet rays, Arch. Dermatol., 100:535-539, 1969.

81. Kaidbey, K., The evaluation of photoallergic contact sensitizers in humans. In Dermatotoxicology (F. M. Marzulli and H. I. Maibach, eds.), Hemisphere Publishing, Washington, D.C., 405-414, 1983.

4

Human Ocular Irritation

William C. Waggoner

Tewksbury Research, Lebanon, New Jersey

Charles J. Minnich and Alan S. Dana, Jr.

Techni-Med Consultants, Inc., Wyncote, Pennsylvania

I. INTRODUCTION

Since the introduction of the rabbit ocular irritation test by Draize et al. in 1944 for evaluating preparations that could come in contact with the human eye, several animal studies have been done in critique or further elaboration and refinement of the Draize method. Further, attempts have been made to explore the interspecies correlation of ocular irritation tests, and also to extend the tests to include human subjects. This present protocol for evaluating preparations in the human eye has been used in clinical testing for over 10 years and encompasses all varieties of cosmetic products and soaps. The method has been used to test the safety of products, and to support product claims of low ocular irritation potential or lack of ocular sting.

II. METHODS

It is important that the product be tested in the use concentration that would most likely come in contact with the human eye. Shampoos and soaps are usually tested in the lather concentration, which is determined by taking lather samples from shampooing or washing from 10 volunteers just prior to rinsing. A quantitative analysis is performed on the lather. The simplest method is to do a loss on drying. This gives an estimate of the average use concentration of the product that might enter the eye. Of course, if the probability is good that the neat product could accidently enter the eye, then, if practical, it is tested in that form.

Data from animal testing are reviewed to ensure risks to the volunteer are virtually absent upon controlled instillation.

Subject volunteers are healthy individuals of either sex with no known history of chronic systemic disease or ocular disorders. Individuals must not have participated in any ocular testing program within the 7 prior days, and must not have had an ocular infection or shown ocular irritation for at least 48 hr prior to instillation of test materials. Subjects are not eligible for participation if they use contact lenses or show ocular irritation at baseline (preinstillation) examination. In addition, subjects cannot have received treatment with sympathomimetic, antihistaminic, or anti-inflammatory medication within the 7-day period prior to instillation of test materials. If necessary, subjects could receive any of the aforementioned medications as treatment after test material instillation.

III. SCORING

Following a thorough medical history and in-depth interrogation, eyes are examined by observing the bulbar and palpebral conjunctivae. The examination can be made by an ophthalmologist or by another trained individual. The volunteer is questioned as to any ocular discomfort or stinging. In addition, the amount of lacrimation is judged. Scores are assessed according to the following subjective scales:

A. Sensorial

0 = None
1 = Minor discomfort
2 = Hurts a little, some sting
3 = Hurts a lot, but tolerable
4 = Severe discomfort (subject requested washout and/or towel)

B. Lacrimation

0 = None
1 = Mild, occasional tesr
2 = Moderate, tears flooding eye, but confined to orbit
3 = Severe, tears flooding eye and leaving orbit

C. Irritation

1. Bulbar Conjunctiva

0 = None
1 = Slightly pink with minimal capillary dilation
2 = Deep pink with definite capillary dilation
3 = Very red with vessels prominently dilated

2. Palpebral Conjunctiva

0 = None
1 = Slightly pinker than normal
2 = Deep pink with definite capillary dilation
3 = Cherry to deep red with vessels prominently dilated

IV. STUDY DESIGN

Various study designs may be used. The basic design has been of a controlled, double-blind, randomized, non-crossover nature. The study is controlled by instillation of one product in one eye and a placebo product, water, or positive control in the contralateral eye. Material for instillation must be sterile and free from particulates. Sterile diluent, sterile containers, and sterile droppers are used throughout for material dilution and instillation. Prior to instillation, containers of material for instillation are maintained at 37-38°C for the duration of instillation. As a rule, 10 subjects are used for each product comparison.

V. TREATMENT

Eyes are graded (using the above scales) prior to instillation of test materials. Then, in accordance with the randomization schedule, one drop (about 0.1 ml) of the appropriate test substance is gently instilled by a trained technician into the cul-de-sac of the assigned eye. Bilateral instillation proceeds as consistently and rapidly as possible, with instillation in the left eye always preceding instillation in the right eye. Tissues are made available to subjects to gently blot the lid area of each eye. Within 10 sec, or as closely as possible following instillation, eyes are examined and scored according to the aforementioned grading scale. Subjects are interrogated for the sensorial component. Examinations are also done at 2 min, 15 min, and 24 hr following instillation. If any residual conjunctivitis is observed at the 24-hr exam, the volunteer is scheduled to return for a 48-hr examination. Any persistent conjunctivitis is treated.

If sufficient discomfort is initially experienced by the volunteer (grade 3-4) upon material instillation (at the subject's election), the specific eye is immediately and gently flooded with isotonic saline, and the maximum value (or score) for ocular sensation in that eye is awarded. The election to rinse with isotonic saline is recorded on the report form. Further clinical evaluations and questioning proceed on schedule. All volunteers are advised at the 15-min examination that they should immediately report any subsequent ocular complications, which might arise prior to the 24-hr examination.

Again, subjects cannot be on any other treatment during their entire test period. Specifically excluded are concomitant sympathomimetics, analgesics, antihistamines, or anti-inflammatory agents unless prescribed as treatment after test material instillation.

VI. RESULTS

At times, marked differences can be observed between products. Usually, any reactions that occur at instillation are completely subsided at the 24-hr examination. Of interest is the interaction between eyes of individual subjects, indicating occurrence of a possible sympathetic effusion. That is, if one eye responds to a test product, the contralateral eye may also begin to respond in absence of a product. Contralateral ocular response is in the form of lacrimation and vessel injection. This finding should be considered when evaluating substances where innocuous controls are used contralaterally.

SUMMARY

This format presents a standardized method for determining the acute ocular irritancy potential in human subjects of certain consumer products in use concentrations. In addition, implementation of this model design can successfully delineate differences in a group of similar products with diverse formulations.

APPENDIX

MEDICAL HISTORY FORM

Date: ________________

Subject No.: __________

Initials:_____ Age:_____ Birthdate:___/___/___ Sex:_____

Race: Caucasian ( ) Black ( ) Hispanic ( )

Other ( )______________________________

Ocular: Glasses (?) ☐ Yes ☐ No Contacts (?) ☐ Yes ☐ No

History of Conjunctivitis (?) ☐ Yes ☐ No

Other______________________________________

Allergies: Hayfever___ Asthma___ Urticaria___ Contact Dermatitis___

Cosmetics___ Medications (type)________________________

Other_______________________________________ ☐ None

Other Significant Medical Problems: Lung Disease___ Diabetes___

Cancer___ High Blood Pressure___ Heart Disease___ Ulcers___

Liver Disease___ Kidney Trouble___ Thyroid Disease___

Rheumatic Disease___ Gall Bladder___

Other (explain)____

____ ☐ None

Are you taking any medication? ☐ No ☐ Yes, detail what medication and why: ____________________________________

____

____

______________________________ ____________
Investigator's Signature Date

APPENDIX 4.1

CLINICAL RECORD

Subject Number_____ Subject's Initials____ Sex: ( ) F ( ) M

Birthdate __/__/__ Age ____ ( ) Right-Handed ( ) Left-Handed

Weight ______(pounds) Height ____ (inches) Eye Color_______

Smoke? ( ) Yes ( ) No

Exam Date	Sensorial	Lachrimation	Bulbar Conjunctiva	Palpebral Conjunctiva
___/___/___ Baseline				
___/___/___ Initial				
___/___/___ 2 min				
___/___/___ 15 min				
___/___/___ 24 Hour				

Comments____

____

____

____

_______________________________ ____/____/____

Investigator's Signature Date

_______________________________ ____/____/____

Technician's Signature Date

APPENDIX 4.2

EXAMPLE

Subject No.______ INFORMED CONSENT

OCULAR IRRITATION STUDY

You have been asked to participate in a 1-day study, which will test a cosmetic product by instillation in the eye in a controlled use fashion. The purpose of this evaluation is to determine if the product could cause a problem to the user when used in a use concentration, if the product should come in contact with the eye. The products will be instilled by a trained technician in the presence of an ophthalmologist. Approximately (#) individuals will participate in this study. Use of any medication, unless first discussed with the study doctor, is prohibited. You must normally use cosmetic products, but not have a history of allergy to cosmetics. Products used in this study will first have completed satisfactory eye testing in animals.

After a medical history is taken by the Investigator or one of the appointees, you will have an initial ocular examination by the doctor at the Clinic. Provided you qualify, you will have one drop of the product instilled into the corner of your eye. You will be asked questions about the feel of the product in your eye, in addition to immediate further examination. You may request immediate eye wash with a simulated tear solution, if you feel the discomfort is not tolerable.

Further examinations and questioning will take place 2 minutes after instillation and 15 minutes after instillation. You will also return to the Clinic for a final examination 24 hours after instillation. If there is residual discomfort or conjunctivitis, you will be scheduled for an examination 48 hours after instillation. Any persistent conjunctivitis will be treated.

You have been selected for this study because you fit the age, sex and other required criteria of the protocol. Your participation in this study will not be of benefit to yourself, but can help determine the potential ocular irritation properties of a particular product. All your records pertaining to this study will be kept confidential by (name of the clinic), and the Sponsor, and could be made available to certain agencies, public or private. If you do participate, you may withdraw at any time without prejudice.

Should you require any treatment for injury or side effects that you think may have occurred as a result of the test treatments, you should contact the Clinic immediately. The Sponsor will compensate you for medical expenses for illness or injury that is CAUSED BY THE TREATMENTS, providing the Investigator had adhered to pertinent provisions of the protocol. Pre-existing medical conditions or disease or conditions not associated with the treatments administered are not covered by this provision. If you

APPENDIX 4.3

experience any difficulties or have any questions during this study please contact the Clinic immediately (phone number).

AUTHORIZATION

I have read and understand this Consent Form, and a copy will be made for me upon my request. The study format has been explained fully to me to my satisfaction and I have been given the opportunity to discuss all aspects of this study and to ask questions. I agree to participate in this study as described, and understand that I am required to complete all phases of the study in order to qualify for payment.

Subject Signature________________________ Phone#__________________

Witness Signature________________________ Date____________________

APPENDIX 4.3 (continued)

5

Assessing Comedogenicity: Current and Future Trends

Otto H. Mills, Jr., and Richard S. Berger

University of Medicine and Dentistry of New Jersey, Robert Wood Johnson Medical School, New Brunswick, New Jersey, and Hill Top Research, Inc., East Brunswick, New Jersey

James P. Bowman and Kim S. Dammers

Hill Top Research, Inc., East Brunswick, New Jersey

I. INTRODUCTION

For years, clinical observations and laboratory studies have pointed to certain extrinsic factors that worsen acne [1]. Reports of acne venenata are well known [2]. In 1972, Kligman and Mills [3] used the term comedogenicity in regard to animal and human model studies. They defined comedogenicity histologically as producing comedones in the external ear canal of the rabbit and subclinical comedones on the backs of human volunteer subjects. A series of reports followed testing for this phenomenon with soaps and shampoos [4] and sunscreens [5]. Additional work in a human model using a noninvasive method called the follicular biopsy compared data from the rabbit and human models [6,7].

The topic of comedogenicity continues to be of interest to both dermatologists and the industries manufacturing products for use on the skin. This chapter will update the work done on comedogenicity. Since several laboratories have published their experiences in the animal model [8-11], and relatively little has been in the human area, this chapter will briefly describe the animal model assay in use in our laboratories and then detail new work done in a human model system. In addition, we will outline work currently underway allowing us to assess comedogenicity on the face. Finally, some future projections for this area will be made.

II. ANIMAL MODEL SYSTEM: METHODS AND OBSERVATIONS

As set forth previously, we continue to use albino male New Zealand rabbits for each test material with 10 applications over a 2-week period [3]. For routine screening, we evaluate comedogenicity using the whole mount sampling technique, and use a scale of 0-4 (0 = none, 1 = trace, 2 = mild, 3 = moderate, 4 = severe).

The whole mount technique [12] (Fig. 5.1) is quicker and less costly than histology [3,13]. In a series of assays done over months with six animals, we saw no evidence that a 4-week assay would be any more helpful than a 2-week study (Table 5.1). In our experience, materials that are comedogenic in this model demonstrate such activity by 2 weeks. We have not seen a 0 rating develop into a positive one (grade of 1 or more) by 4 weeks.

We do, however, use histology when we wish to further view the follicular canal and its epithelium and assess such characteristics as acanthosis and hyperplasia. When histology is performed, horizontal cuts are made through tissue taken using an 8-mm punch biopsy. Serial sections are done and stained with hematocylin and eosin. Judgments for hyperkeratinization are made on sections from the middle and lower parts of the follicles.

It is important to note that it has not been possible for us to correlate clinical observations of dilation of the follicles of the rabbit ear with the whole mount tissues taken from the same area (Table 5.1). Review of many assays over the last few years point to frequent positive readings based on clinical observations which could not be supported by whole mount retrieval of hyperkeratinous material from the follicles. Apparently some compounds cause surface changes that appear to resemble plugging when in fact there is no plugging in the follicle. Therefore, it is very important that sampling in this assay is either done using the whole mount or histology techniques.

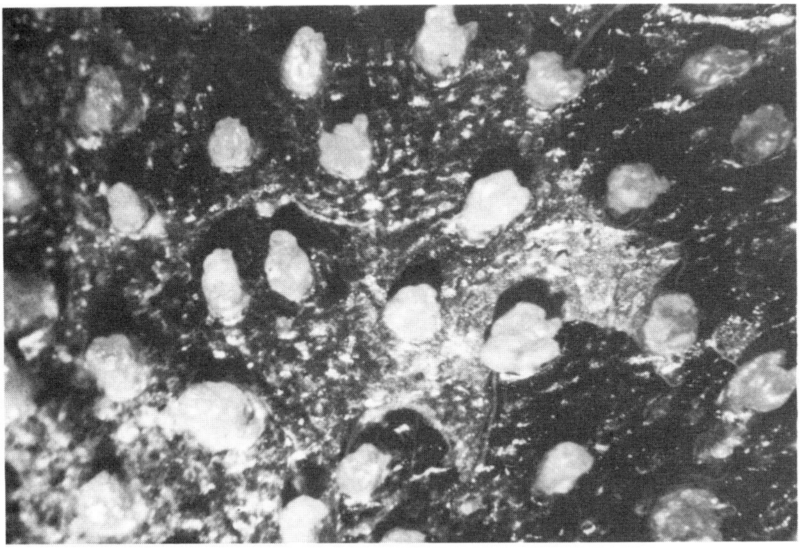

FIGURE 5.1 Whole mount sample of rabbit ear, as seen through the stereomicroscope (50×).

TABLE 5.1 Rabbit Ear Assay

Product/ material	Clinical observations		Whole mount evaluations	
	2 wk	4 wk	2 wk	4 wk
A	2.6	3.0	0.3	0.5
B	1.3	2.0	0.2	0.0
C	1.7	2.3	0.3	0.7
D	2.0	3.7	0.0	0.3
E	2.3	2.3	0.0	0.2
F	0.3	1.3	0.2	0.3
G		2.6		2.5
H	2.7	3.0	2.0	2.2
I	1.7	2.0	2.0	1.7
J	1.0	1.0	1.7	1.1
K	3.3		3.7	

TABLE 5.1 (continued)

Product/ material	Clinical observations		Whole mount evaluations	
	2 wk	4 wk	2 wk	4 wk
L	0.7	0.5	0.7	0.7
M	1.0	1.0	0.5	0.3
N	1.7	2.0	2.0	1.7
O	1.0	1.0	1.7	1.2
P	1.0	1.3	0.0	0.0
Q	0.3	0.0	0.0	0.0
R	0.7	1.3	0.0	0.0
S	3.0	3.0	0.0	0.0
T	1.3	2.0	0.2	0.0
U	2.7	2.0	0.5	0.3
V	1.3	2.7	0.3	0.3
W	2.0	2.3	0.7	0.5
X	3.3	3.0	0.0	0.0
Y		1.6		2.5
Z		1.1		0.3
AA		2.0		0.3
BB		2.5		1.7
CC	3.0	2.0	1.0	1.3

III. HUMAN MODEL SYSTEM: METHODS AND OBSERVATIONS

The human model for assessing comedogenic substances that was previously described [6] used adult black male volunteer panelists who exhibited prominent follicular orifices and a preexisting tendency to form microcomedones, as determined by the follicular biopsy technique. The test substances were applied under occlusion for 1 month to the upper part of the back. The degree of follicular hyperkeratosis was reassessed by the noninvasive follicular biopsy technique, which employed a fast-setting cyanoacrylate glue to remove the follicular contents. Recently, we set out to compare the model system when black males as well as white females were empanelled as test subjects.

Approximately 12 adult black males and 12 adult white women were enrolled in each study. Selection of suitable subjects was done utilizing the follicular biopsy technique [7]. This technique consisted of coating the skin with a thin layer of methyl cyanolacrylate glue and pressing a glass microscopic slide to the skin, which spreads the polymer evenly. This adhesive sets in approximately 1 min, after which the slide is peeled off, bringing with it the follicular horny extensions (microcomedones). The specimen is examined under the stereomicroscope and subjects who did not show some microcomedones are rejected.

Four 4-cm square areas are mapped out above the scapulae on each subject. One of these sites serves as an untreated control. The test substances are liberally applied to squares of nonwoven cotton (Webril). The nonwoven cotton is then fastened to the skin by strips of tape. All test materials are coded and randomly assigned to each test area using block randomization. Applications are made on Mondays, Wednesdays, and Fridays for 4 consecutive weeks, for 28 days of continuous occlusive exposure. At the end of this time, each site is then sampled by the follicular biopsy technique and examined under the stereomicroscope. The test and control sites are each assigned a value, as indicated below. A positive response for the test materials is indicated by the presence of microcomedones greater than those present in the untreated control. Comedogenic potential is assessed on a 4-point scale:

0 = Noncomedogenic (similar to the control)
1 = Small microcomedones
2 = Moderately sized microcomedones over most of the field
3 = Large, globoid microcomedones over the entire field

Each test site grade is compared with the untreated control for an individual subject. The grades of the entire panel are then averaged to give a mean rating of comedogenicity. Based on the nature of the grading scale, one can use binomial statistical analysis if desired. Table 5.2 shows the means for several comparative assays (N = 115) done over the course of 1 year.

The data generated from this series of comparative assays in the human indicate that good correlation exists between microcomedone formation on the back of black males and white females. In addition, the latter population will occasionally

TABLE 5.2 Human Model (N = 115)

Test material	Black males	White females
I	0.32	0.35
II	0.46	0.10
III	1.32	1.50
IV	0.35	0.25
V	0.0	0.08
VI	0.19	0.28
VII	0.19	0.17
VIII	0.31	0.35
IX	0.35	0.25
X	0.0	0.08
XI	0.19	0.28
XII	0.19	0.17
XIII	0.31	0.35
XIV	0.21	0.29
XV	0.04	0.17
XVI	0.21	0.17
XVII	0.67	0.42
XVIII	0.08	0.12
XIX	0.08	0.10
XX	0.08	0.10
XXI[a]		

[a]Follicular biopsy sample not taken because of irritation caused by test material.

demonstrate papules and pustules in the test sites (Fig. 5.2). In the original report of Mills and Kligman [6], this was considered a drawback; however, this development may actually be helpful in identifying materials which could be pustulogenic in nature, and particularly those that might only be pustulogenic not comedogenic. Another new feature of this human model assay is that 12 panelists allows for block randomization of the three test materials and one untreated control.

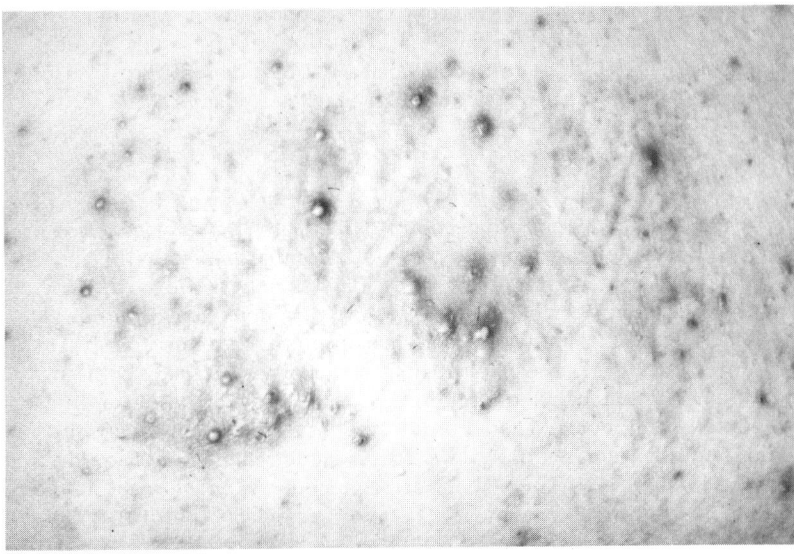

FIGURE 5.2 Papulopustular lesions on the back of white female panelist after fourth day of human model assay.

As comedogenicity is a phenomenon which appears to impact on acne-prone skin, it is important to use acne-prone subjects. Most important in the human assay work is to screen for the presence of some microcomedones in each panelist.

IV. FUTURE TRENDS: FACIAL ACNEGENIC ASSAY

Currently we are gathering data on a facial acnegenic assay. We have reported trials using the follicular biopsy on the face to screen for comedogenicity [14,15]. Utilizing this noninvasive methodology and counting clinically evident acne lesions (noninflammatory and inflammatory), comedogenic testing on the face has become part of our in-use cosmetics testing. By including a certain percent of acne-prone individuals in a test group, information can be gained on comedogenicity, pustulogenicity, and acne-aggravating potential. This assay allows for evaluation of the final formulation of a single product or for product combinations or regimen testing. It is important to know how individual products intended for use together will influence the skin in regard to these phenomena. Another advantage of this method is that safety information, such as clinical signs of irritation and symptoms of burning, stinging, itching, or tightness, is evaluated during these assays.

V. SUMMARY

The more we study comedogenicity, the more it becomes clear to us that the evaluation of the final formulation is critical. Many materials will test positive at 100% concentration, but quickly lose their comedogenicity when the concentration is lowered. Final formulations consist of numerous ingredients at varying concentrations interacting chemically. It may even be possible that noncomedogenic ingredients might produce a final formulation which is comedogenic. As stated some years ago, "What is ultimately important is the comedogenicity of the finished product." Giving a cosmetic chemist or an acne patient a list of individual ingredients to avoid is rarely helpful. Such a list restricts one's choices of products unnecessarily. Should a final formulation test positive or comedogenic, what can be done? One can change percent concentration of known or suspected comedogenic ingredients, and if this does not work, the active chemical may have to be eliminated from the formula.

Reviewing our work in animal and human models and our recent experience using the facial acnegenic assay, it becomes clear that the latter represents the test which offers the most accurate and useful information, as it predicts both acnegenic and acne-aggravating potential.

REFERENCES

1. Mills, O. H., and Kligman, A. M., External factors aggravating acne. Dermatol. Clin., 1:365, 1983.
2. Plewig, G., and Kligman, A. M., Acne Morphogenesis and Treatment. Springer-Verlag, Berlin, 1985.
3. Kligman, A. M., and Mills, O. H., Acne cosmetica, Arch. Dermatol., 106:843, 1972.
4. Mills, O. H., and Kligman, A. M., Acne detergicans, Arch. Dermatol., 111:63, 1975.
5. Mills, O. H., and Kligman, A. M., Comedogenicity of sunscreens, Arch. Dermatol., 118:221, 1978.

6. Mills, O. H., and Kligman, A. M., A human model for assessing comedogenic substances, Arch. Dermatol., 118: 903, 1982.

7. Mills, O. H., and Kligman, A. M., The follicular biopsy, Dermatologica, 167:57, 1983.

8. Morris, W. E., and Kwan, S. C., Use of the rabbit ear model in evaluating the comedogenic potential of cosmetic ingredients, J. Soc. Cosmet. Chem., 34:215, 1970.

9. Fulton, E., Bradley, S., Aquendex, A., and Black, T., Noncomedogenic cosmetics, Cutis, 17:344, 1976.

10. Hambrick, G. W., and Blank, H., A microanatomical study of the response of the pilosebaceous apparatus of the rabbit's ear canal, J. Invest. Dermatol., 26:185, 1956.

11. Kanaar, P., Follicular properties of fatty acids in the external ear canal of the rabbit, Dermatologica, 142:14, 1971.

12. Kligman, A. M., and Kwong, T., An improved rabbit ear model for assessing comedogenic substances, Br. J. Dermatol., 100:152, 1963.

13. Kligman, A. M., Pathogenesis of acne vulgaris. II. Histopathology of comedones induced in the rabbit ear by human sebum, Arch. Dermatol., 98:58, 1968.

14. Berger, R. S., Mills, O. H., Stephens, T. J., and Hair, J., The incidence of facial acne in adult females. Clinical Studies Session, American Academy of Dermatology, 3:9, Boston, 1986.

15. Mills, O. H., Berger, R. S., Stephens, T., Drake, K., and Fisher, L., Assessing acnegenic and acne aggravating potential. American Academy of Dermatology, 46th Annual Meeting, 1987, p. 108.

6

Protocols and Strategies for Biophysical Testing

Harry R. Elden

Xienta, Inc., Bernville, Pennsylvania

I. INTRODUCTION

The cosmetic product industry is being challenged by two activities; namely, *rising claims* of product efficacy, and concerns that *claims exceed* substantiation. An outcome of these challenges is a call to the industry to improve specification of products and to substantiate claims. If it is necessary to test products more extensively, there also is a need to focus product efficacy. Thus, a basic need exists to identify features of skin that ought to be addressed by ingredients, formulations, and treatment strategies.

It is not unreasonable to expect products to satisfy consumer needs, especially if they are claimed to do so. But, what are the real needs of skin in a heterogeneous consumer market. Real physiological needs of human skin must be identified before ways to satisfy these needs can be developed. Claims then can be made to satisfy needs with expectation that consumers and regulatory affairs will be satisfied.

It is proposed that human subjects be studied in order to identify skin care needs, and formulated products be tested on human subjects. The first and foremost property of interest is that all consumers are age dependent. Then, one has to recognize the great diversity of body build, skin type, and biochemical milieu. In addition to aging, therefore, general heterogeneity of consumers is the task to be addressed.

The life expectancy of consumers is rising steadily, and the number of people living to older age also is increasing. The key word here is *living*, rather than just surviving. An increasing number of mature adults know that vitality of youthfulness endures and can be retained as they grow in knowledge, judgment, and affluence. Their knowledge, discretion, and affluence drive the worldwide skin care economic enterprise. However, rising consumer needs call for more assertive products and treatment strategies. While rising consumer need is a major factor in increasing claims, it challenges the industry.

The topic to be faced is that market-driven claims often exceed demonstrated efficacy. Product claims are not substantiated adequately, and according to the U.S. Food and Drug Administration (FDA), claims far exceed the knowledge base pertaining to implied efficacy and mechanisms of action [1]. This is a major challenge! But, it also is an opportunity for cosmetic chemists, skin scientists, and testing laboratories to advance cosmetic science by providing the needed information. This article, therefore, deals with the issue of claims versus substantiation by proposing strategies and protocols to assess aging human skin.

II. SUBJECT SELECTION

The American population is a heterogeneous mix of ethnic, cultural, economic affluence, and age groups. Sharply delineated communities are interspersed throughout the nation. Skin characteristics tend to be similar within ethnic groups but differ markedly across ethnic groups. Communities share common physiological experiences; i.e., aging, sun exposure, seasonal dry skin, oil condition, acne, and so forth. But, for the most part, cosmetic product formulators do not know the full range of skin types that exist in the U.S. market. What is worse is that mass marketing strategies assume that all skin types are the same, since claims or directions for use do not specify variances for skin types. It is important, therefore, to emphasize the need to determine the varieties of skin types making up the consumer market. Without this information, one cannot identify product types needed to satisfy individual consumer groups.

Thus, it is valuable to consider ways to study types of skin that make up the consumer group to be served by cosmetic product formulators, manufacturers, and market strategists.

Features of consumers' skin can be determined during recruitment of subjects for product testing panels. Age and sex are obviously important properties, but they are underutilized in describing samples of populations. Age per se is an inadequate descriptor; functional age, however, provides a better indication of vitality. Gender can be ascertained, but refinements in the male versus female character exist in all subjects. It is proposed, therefore, that functional age be used as a minimal descriptor of subjects. Methods for determining functional age, softness, smoothness, roughness, dryness, oiliness, and skin androgenicity will be discussed.

A. Strategy

Panelists can serve two purposes concurrently. They can be used to study product activity during panel testing, but they also can indicate features of consumer groups: (1) Tests used to screen panelists are a resource, providing sufficient measurements are made. (2) Baseline data used for product testing are just as valuable as experimental data. (3) Patrons, when their skin condition is evaluated by estheticians, massage therapists, and dermatological nurses can show what consumers need. Fortunately, a number of noninvasive methods can be used to study aging human subjects. Properties that depend upon age are attended with the goal of subclassifying patrons according to skin type and androgenicity.

B. Functional Age

It is widely held that the rate of aging depends upon individual propensity. A well-studied counterpart to this idea is pre-adolescent growth and development, which can be measured easily. Adult aging, however, does not begin sharply. It commences slowly, is not localized, and involves the entire hierarchy of physiology.

Functional age of physiological subsystems has been studied intensively in the gerontological literature. Ordinarily, systems are selected for their measurability and sensitivity. It is possible to use any age-dependent property to estimate functional age, providing one has a large enough sample for reference.

A study by Borkan and Norris [2] is particularly relevant, because it addresses the topic of the appearance of agedness. These authors used 24 properties of age-dependent systems to estimate rate of aging. Subjects were obtained from the Longitudinal Study Group, Gerontology Research Center, Baltimore; 1086 male subjects were examined for blood, pulmonary, circulatory, cognitive and psychomotor performances. Performance scores for individuals were compared with mean values for the whole sample. Excursions above and below mean values were assigned degrees of agedness, based upon age regressions for each parameter.

Some subjects were found to be aging faster than others, whereas others were aging slower. An interesting application was described by comparing the visual impressions of agedness with actual functional age for subjects that were selected from the same chronological age. Normalized scores of certain subsystems correlated with visual estimates of perceived oldness. They also showed that mortality for equal-age subjects correlated with functional age. Higher death rates were found for persons whose functional age exceeded their chronological age. Estimation of agedness was done by clinical evaluation of body surface. All the visual signs of aging skin that we know about were present, plus those that so far have eluded physical measurement.

Objective functional age measurement of body systems correlated with the clinical (esthetic) estimate of perceived oldness or youngness. Functional age as measured by objective laboratory tests, therefore, predicts or correlates with clinical perceptions of agedness (skin).

A study by Hollingsworth et al. [3] showed that rheology of skin could be included in the test battery. Seventeen properties were measured on 437 subjects; 169 males and 268 females (Japanese). Refinements during analysis decreased the test battery to nine tests, one of which was skin elasticity. Elasticity was defined in terms of rheology; i.e., time required for a lifted-pinch of skin to flow back to conformity (back of hand).

Recovery time was instantaneous below 25 years of age for males and females. It increased to about a second at age 35 (males) and 45 (females). Thereafter it increased exponentially to over 50 sec for males and above 40 sec for females at 60 years of age. The authors broke out the data for males and females in 10-year intervals for the following tests: Skin elasticity, systolic blood pressure, vital capacity, hand grip-strength, light extinction, vibrometer, visual acuity, audiometry, and serum cholesterol. Correlation coefficients were computed for interaction of test parameters with age and each other. Functional age was estimated by each test and compared with others in the test battery. An important observation was made by Hollingsworth et al. [3]; i.e., it is after age 40 years that *life begins to end.* Below age 40, there is little variation in these properties with aging.

A study by Furukawa et al. [4] used functional age to investigate circulatory properties of normal and hypertensive subjects. A screening study using 53 normal subjects (26-70 years of age) showed that organ function tests do not depend strongly upon age. An additional screening study of 111 subjects 21-83 years of age measured cutaneous sensitivity to vibration, plus ocular accommodation and vital capacity. These systems increased test sensitivity for functional age. The third screening study examined 34 males (21-74 years) and 31 females (27-73 years) using the second set of test parameters. Hypertensive subjects exhibited greater functional age than normal tensive subjects. The fourth screening study, done on 110 subjects between 21 and 71 years of age, used the Master's exercise test. Fitness features detected by this test battery also revealed functional age differences between hyper- and normotensive subjects.

Review of the literature by these authors shows that functional age can be estimated by standard laboratory tests of human performance. Systems that depend strongly upon circulation or oxygenation seem to have greater age dependence than others; i.e., cutaneous vibratory sensitivity of skin involves neurosensory receptors that depend strongly upon oxygen. Tests, however, generally are noninvasive and use calibrated and standard instruments.

A commercially available test station can estimate functional age reasonably well, but tests must be detected by electronic devices. Hochschild's [5] description of this system confirms reports by other leading investigators. Functional tests reflect physiological subsystems as well as human individuality overall. There probably is a selection of tests that reflect skin physiological age better than other tests, but they have yet to be discovered. One purpose of this article is to suggest what they are and to relate them to overall human functions.

Skinfold thickness has been studied extensively and used to estimate fat content of the body. Standard calipers and specified sites are all that is required to measure human individuality. Durnin and Womersley [6] studied skinfold thickness and buoyancy of 209 males and 272 females between 16 and 72 years of age. Mean ponderal index for these subjects is 2.43 and 2.25 for males and females, respectively. Skinfold thickness increased with aging, but inconsistently, depending upon where it was measured; i.e., biceps, triceps, scapular, and iliac crest. It is interesting to note that females had thicker skin than males. The authors also determined body density by measuring buoyancy in water; body fat was computed from body density.

Arithmetical analyses of data showed that body density decreased linearly with log (total skinfold thickness). Body fat of females was higher than that of males, whereas it increased with aging in both sexes. Various manipulations of skinfold thickness obtained at the four sites were used to estimate body density. Generally, computed density agreed with measured density.

This study shows that skinfold thickness can be measured easily and reliably. Data can be used to estimate body density and body fat content. The authors draw attention to the question of proportion of fat in skin versus that internally disposed. Studies in the literature do not agree, but they suggest that about 50% of fat is found subcutaneously. Findings support the claim that human subjects can be categorized according to age, gender, height, weight, ponderal index (weight/height), skinfold thickness, and density. Variations in measurements imply differences in functional age of these systems.

Another test parameter for functional age analysis of skin is its compressibility. While Durnin and Womersley [6] suggested that compressibility decreased with aging, Brozek and Kinzey [7] measured it on 107 male subjects, 20-69 years of age. They observed the amount of compression attending increased caliper pressure from 5 to 10 gm/mm^2. Compressibility decreased with aging beyond 30 years, but responses depended upon sites (arm, calf, scapular, and waist). A kinetic analysis of compressibility was studied by Kirk and Chieffi [8], who applied an indenter of known weight; indentation was measured at time periods following loading. Elasticity was defined in terms of skin movement; i.e., (1) immediate indentation, (2) 2-min indentation, (3) immediate rebound, and (4) 3-min rebound. Studies done on 263 women from 18 to 104 years showed that all measured parameters decreased with advancing age.

These early biophysical measurements of aging human skin provide an insight into present-day methods that use advanced engineering constructions and computerized data acquisition.

The earlier studies are valuable today, showing that simple methods can be just as effective as more elegant devices. And they provide estimates of population behavior based upon samples that approximate 50 to over 100 subjects. This is important to keep in mind. If one is asking skin scientists to increase the amount of testing done on panelists, anything that decreases time and cost will likely improve acceptance of this suggestion.

Simpler test methods are necessary!

A rheological test method that is elegant for its simplicity is that used by Tosti et al. [9], where amplitude of a bouncing indenter is recorded continuously. Energy returned to the indenter relative to that imparted to skin was computed as ratio of amplitudes. Rebound energy decreased linearly with advancing age (8-80 years) for 46 subjects; forehead, hand, and thigh. Traces of amplitude versus time showed patterns unique to skin conditions. Sun-exposed skin showed greater loss of rebound energy than nonexposed skin. Pathological skin conditions also changed the pattern markedly.

A comprehensive description of the rheology of aging human skin is that reported by Leveque et al. [10] for 150 subjects from 8 to 89 years of age. Elasticity, cutaneous relief, transepidermal water loss, desquamation, and photoplethysmography were measured. All physical properties changed with aging. Similarly, torsional elasticity of aging human skin was studied by Sanders [11] for 19 subjects from 6 to 61 years of age. Female skin exhibited greater elastic stiffness than did males, whereas elastic modulus increased with age for both sexes. Another simple method to measure elastic rheology of aging human skin is the diaphragm method of Grahame and Holt [12], based upon Tregear's [13] earlier description.

The above reports were selected to emphasize an important point. It is not necessary to use highly esoteric methods, instruments, or study designs to obtain relevant information on aging skin. The simpler methods are most likely to serve the purposes of cosmetic chemistry because they are less costly, efficient, and rapid. Functional age of skin can be estimated by any of these methods because properties measured are age dependent. Excursions of individual values from group means indicates faster or slower progress of skin rheology with respect to aging. The methods are noninvasive and not traumatic. They can be administered by skilled technicians, but overall processing is helped by computerization of data collection.

The optical properties of skin provide noninvasive access to the stratum corneum, epidermis, and dermis. Measurements can be used to categorize skin according to composition, structure, and geometry. These aspects of skin optics have been reviewed by Anderson and Parrish [14]. Generally, black and white skin absorbs 93-95% of incident radiation between 250 and 3000 nm wavelength. This large amount of energy is absorbed and scattered within layers. Scattering is due to the size and shape of heterogeneous structures, whereas absorption is due to amino acids, nucleic acids, and pigments. Skin also has a natural fluorescence. The broad absorption band of skin is due to tryptophan and tyrosine (275 nm), nucleic acids (260 nm), and urocanic acid (277 nm); the absorption band is centered at 275 nm. Epidermal fluorescence that is emitted between 330 and 360 nm is excited at 280 nm, consistent with tryptophan and tyrosine fluorescence.

Melanin is not a neutral density filter because absorption increases steadily toward shorter wavelengths (1200-250 nm). Absorption by melanin beyond 1100 nm (near infrared) is negligible. Dispersal of melanin in Caucasian skin is believed to protect better against ultraviolet (UV) damage than discrete melanosomal containment of the same amount of pigment. And optical differences between Caucasian and Negroid skin are believed to be due more to the pattern of melanin dispersion than to optical transmission of the stratum corneum; a 10- to 30-fold UV sensitivity variation versus threefold transmission of variation, according to Kaidbey et al. [15]. Urocanic acid is believed to be a natural sunscreen for skin, and it can be extracted nondestructively from human skin. Further, urocanic acid can be deposited on the skin surface by sweating [16].

It is interesting to note further that the ratio of diffuse/direct transmission is not wavelength dependent for UV or visible light. This suggests that refractive index heterogeneities account for scattering more so than does irregular particles. Only about 5% of collimated incident radiation in the 350-3000-nm region is scattered within Caucasian epidermis.

The optical properties of dermis differ markedly from that of epidermis. Scattering is of greater importance owing to major fibrous proteins and blood vessels. Consequently, light penetrates the dermis better at longer wavelengths (near IR) than in the near UV. Bloodborne pigments of reduced and oxyhemoglobin, bilirubin, and beta-carotene absorb energy in the dermis. Penetration through the dermis is such that 1% of incident radiation (605-805 nm) penetrates the entire chest wall; penetration at less than 550 nm is less than 10^{-5}.

The above synopsis of Anderson and Parrish [13] strongly encourages using optical properties to categorize human skin. Measurements can be made through the ears, webs (toes and fingers), and perhaps nasal and cheek skin. Variations in pigmentation, vascularity, and dietary input of population subgroups might be revealed by the optical features of skin.

A study reported by Leveque et al. [17] supports this suggestion. Reflection of incident light was measured above the skin surface to avoid blanching owing to contact-compression, and a color index was computed. Ratios of intensity of reflected light determined post- versus pretreatment were used to study vasoactive agents, an oil in water emulsion vehicle, and erythema. A typical result showed that the color index increased with the minimum effective dose (MED) used to induce erythema. Instrumental assessment was parallel with clinical estimation, but sensitivity was greater.

Dawson et al. [18] proposed a semitheoretical model of skin surfaces that reflect incident light. The reflectance of skin was compared with that of white tile at wavelengths from 450 to 700 nm. Curves sharply delineated dark Negroid, light Negroid, and Caucasian skin. Erythema and melanin pigmentation could be differentiated. The portable instrument, semitheoretical model and data of this system are ideally suited to categorize human skin in a working situation.

It also is noted that the refractive index of human skin can be measured routinely using ellipsometric methods, according to Pugliese and Milligan [19]. These authors showed that the refractive index of human skin increased linearly with aging, according to the equation,

$$n = 0.0054 \times \text{age} + 0.633$$

where n is the real component of the refractive index and age is in years. Dry skin had a higher refractive index than did moist skin. And application of topical moisturizers changed the refractive index.

A paper by Solan and Laden [20] directly relates the optical properties of skin to cosmetic chemistry. Optical transmission spectra were measured for isolated sheets of stratum corneum immersed in various liquids. Hydration increased transmission at all wavelengths, both vapor and liquid. The most interesting finding is that transmission *increased* markedly and progressively when the refractive index of liquid was raised to match that of the stratum corneum. Maximum transmission occurred at a refractive index of 1.55, that approximating skin.

A number of newer methods also can show dependence on age of the biophysical properties of skin. Profilometry by Cook [21] is valuable because it uses replicas which can be made in the field and submitted to a central laboratory for analysis. Surface contours are recorded and then analyzed by geometric and harmonic methods. Amplitude versus time plots are described empirically and theoretically, yielding characteristics of surface roughness commonly used by engineers. Findings show that surface depends upon age, site, direction, flexion, and cosmetic treatment. Fifteen parameters were used for surface analysis. Other instruments and methods have been proposed and used. Hoppe [22] used four parameters, each of which indicated increased roughness with aging according to a first-order equation.

This search through methodologies is guided by one objective; i.e., find methods to reveal the age dependence of the skin surface for ethnic groups, male and female subjects, and different endocrine states. A study by Corcuff et al. [23] clearly illustrates a method that achieves this goal.

Using surface replicas again, photometric illumination provided shadow patterns that revealed depth, spacing, and orientation of skin lines (inverse images). Two crease lines were noted on the volar forearm. One line disappeared with aging, as shown by an increasing percentage of people missing the line. The remaining principal line became oriented along the long axis of the arm; however, there was considerable exception in older subjects. Density of lines did not change markedly, but depth of furrows increased. Dispersion of data suggested variations in the functional age of subjects.

These selected studies show that human skin is multifaceted, even though only a few sites have been studied. It is very likely that further biophysical studies will only emphasize the uniqueness of skin in individuals and throughout the population. There are more than enough mechanical, electrical, optical, and spectral properties to classify skin. What is needed is refinement of these methods, reduction to simplicity, and application to the task.

C. Skin Typing

It is generally known that skin can be classified into six types (I-VI) according to solar sensitivity. Skin therapists (estheticians, massage therapists, and dermatological nurses) would add oily/greasy, dry/scaly, and rough/smooth conditions to the list of skin types; soft or edematous features also may be added to the clinical appraisal of nondiseased skin. Even so, these conditions do not completely categorize human skin with respect to ethnic, age, and sex groups.

Skin color, for example, is an important feature that can be measured nonvasively and quantitatively. Van Oort et al. [24] used a colorimeter to compare the skin of different parts of the body and among different individuals. Mean and spread of color values in a given population were measured with a spectrophotometer. Physical color values were transformed to perceptual coordinates of luminance and chromaticity of uniform color space; i.e., L (brightness), u (redness [+] and greenness [-]), v (yellowness [+] and blueness [-]). One hundred Caucasian (Netherlands) subjects (8-76 years of age) were measured on the palms, forearm, and cheek.

Skin color measurements showed approximately normal statistical distribution (palms, forearm, and cheek). Only for palm and cheek is there a significant difference between males and females in u. L and v did not differ significantly. L, u, and v parameters also did not change with aging. Factor analysis was used to reveal three main groupings of data. The authors concluded that precision of skin color measurements is inversely related to color saturation. Variations in test parameters with sites were described in terms of optical properties, circulation, melanin pigmentation, and carotene content and skin thickness. It was found that 95% of the subjects were within a width of four standard deviations; absolute values were $L = 61.6 \pm 1.4$, $u = 26.7 \pm 0.9$, and $v = 26.9 \pm 0.4$.

The following statement of the authors relates this study to purposes of this review article [24]:

> Climatic, physiological and ethnical factors will probably make it necessary to modify these color data, before it will be applicable in the artificial cosmetic replication of facial skin.

Physiological condition of skin was used by Mochizuka et al. [25] to type human subjects. Measurements were made on 500 female (Japanese) subjects who rated their skin as oily, dry, or normal. Clinical self-ratings were compared with objective data obtained from sebum quantity, surface microtopography, skin pH, transepidermal water loss, and buffering capacity. Twenty percent classified their skin as oily, 50% normal, and 34% dry. The ratio of numbers of dry/oily increased with aging. The percentage of smooth, intermediate, and rough skin was 12, 35, and 53%. The percentage of rough skin decreased with aging (17-60 years), whereas the intermediate condition increased. Sebum production and water loss seemed to be normally distributed among the subjects. Simple and multiple correlations were done among the measurements. Findings revealed a complex pattern of interactions. The authors proposed *nine categories* of parameters based just upon sebum production, surface roughness, and self-ratings. Review of all data by the authors suggested *three groups* of subjects: *oily skin* type (large sebum production and rough surface); *dry skin* type (low sebum production and smooth surface); and *normal skin* type (50% of subjects). Skin surface pH and buffering capacity were distributed independently of sebum production and roughness.

This study also showed the multidimensional nature of skin functionality within a close ethnic group that used only forehead samples and one sex (females). Obviously, the usual terms of oily, dry, normal, smooth, rough, etc., are a vastly simplified vocabulary for the real condition of skin.

A similar study of skin type was reported by Hartzshtark and Dikstein [26] with 500 subjects; 75% (20-39 years), 25% (40-69 years), and 38 subjects unclassified. About 80% of the group presented a cosmetically dry skin condition; i.e., no sebaceous pores on cheeks, flaky skin, cracked skin or dry-feeling skin. The authors used only two biophysical tests; i.e., indentation-rebound and skin pH. It was possible to categorize subjects on the basis of indentation (penetration and rebound) and pH with respect to age and dryness.

Measurements were classified as follows: I = elevated indentation only; EL = low elasticity only; B = I and EL in the same person; TI = total cases with elevated indentation; TEL = total cases with low elasticity; and T = total cases revealed by indentometry. Measurements were also grouped according to age and skin pH. Findings again show complex interactions, but generalizations do emerge. Elevated indentation and low elasticity are associated with increasing age; elevated indentation is associated with dry skin. And elevated skin pH indicates damaged skin in the older women and predicts damaged skin for younger women. The authors [27] also reported that 93.7% of their patrons were pleased with the results of cosmetic skin treatments recommended by these measurements.

It is clear, therefore, that biophysical measurements can be used to classify human skin. The number of categories, however, is likely to be very great. Measurements can be used to predict responses to treatments. It is reasonable to expect that further use of these strategies will continue to reveal the needs of aging skin that could be addressed by cosmetic formulations.

D. Seasonal and Circadian Rhythms

Seasonal variations occur in skin conditions, as measured by biophysical methods. Potts and Burras [28] measured viscosity of male skin, during summer and winter months, for two age groups: 20-39 years and 40-65 years. Velocity of propagation and damping distance were determined between 100 and 1000 Hz as vibration was applied lightly to the surface. Surface layers responded rapidly to immersion in water. And the stratum corneum contained more water in the summer than in the winter. Low-frequency viscosity of young subjects' skin was greater than that of older subjects for both seasons of wet and dry conditions. These findings suggested that aging decreased hydration of stratum corneum, confirming the earlier work of Potts et al. [29].

Electrical impedance of human skin measured by Tagami [30] also showed seasonal variations in addition to a rapid response to immersion. The summer months increased the hydration of the stratum corneum by 10 times compared with the winter months. Increased hydration of summer skin also was confirmed by Manschot [31]. Extensibility was nearly doubled in summer compared with winter. Purely elastic deformation, however, did not change with seasons. When initially dry inextensible skin of the winter months was hydrated extensibility increased rapidly and substantially. The author suggested that overall extensibility was limited by that of the stratum corneum because responses were rapid.

It is tempting to accept the suggestion that dehydration of the stratum corneum by seasonal low humidity per se is sufficient to account for induction of dry skin. Even though topically applied water immediately softens the surface, according to the above dynamic mechanical and electrical data, water also evaporates rapidly from seasonally damaged (dry) skin. Water alone does not improve the condition of dry skin. Even occlusion of dry skin by petrolatum, leading to minimal dryness in 21 days, regresses to the original condition when occlusion is removed [32]. Obviously fundamental changes in skin take place during seasonal changes of human systems. Subjects who are susceptible to seasonal dry skin might be a unique group within the ever-widening field of skin types. Effective alleviation of seasonal dry skin probably will require more than just occlusive products with improved sensory attributes. A direct attack on rhythmic dynamics of skin is needed.

Seasonal susceptibility of skin to the environment leads to monthly and circadian rhythms of skin. If there are marked changes in the physiology of dermal systems at monthly, daily, or other periods, a need exists then to screen subjects on this basis for testing and treatment. Alternatively, examination of rhythmicity in skin function could generate new product concepts based upon timing of product application.

Monthly variations in skin physiology were reported by Stephenson and Kolka [33]. Recall that females have a menstrual temperature pattern in which (esophageal) core temperature is about 0.5°C higher during the luteal phase than during the follicle phase. Extensive studies show that threshold temperatures for sudomotor and vasomotor activities also vary (cricadian) for males and females. The authors showed that forearm blood flow and onset of sweating occurred at higher temperatures in the PM compared with the AM for both luteal and follicle phase. Forearm blood flow and sweating rate were substantially reduced in the luteal compared with the follicle phase, depending upon the level of core temperature (exercise). The threshold for initiation of sweating was higher in women than in men. Plasma volume shows a peak around ovulation and 10-13 days postovulation, and a minimum around day 2 and again at day 20, according to Turner and Fortney [34]. The following comments by the authors support the purpose of this article: "Menstrual cycle phase and time of day are as important as fitness level, aerobic power and age, when studying thermoregulation in women."

Since skin temperature is a major factor in most biophysical properties of skin, circadian and menstrual timing of subjects are important factors. Again, one could use these times to improve assertive skin treatments.

Further consideration along these lines would only sustain points that seem well founded; i.e., esthetic features of human aging skin are multidimensional; noninvasive methods of basic physiology, physics, and physical chemistry reveal unique clusters of properties. Features seem to settle around oily, dry, rough, stiff, and wrinkled attributes. Even though this seems clear, it is worthy to also notice that body surface has not been studied. And there are a number of well-studied physiological systems in applied physiology and gerontology that relate to skin assessment; these too need to be added to the skin science of cosmetic chemistry.

III. BODY HAIR PATTERNS

Human hair is a unique remnant of an earlier time when body cover was necessary. Today, body hair is a vestigial appendage; i.e., it is disappearing slowly from the skin. Speculations about the decreasing purpose of body hair lead to insights for understanding the increasing role of skin in body welfare and esthetics.

Body hair is believed to have helped maintain body temperature. During hot seasons, body hair reflected UV sunlight to decrease energy absorption by skin. And during cold seasons, thermal insulation by body hair conserved body heat. Hair also permitted skin to slide smoothly over rough srufaces, and perhaps to deflect sharp objects that would have penetrated the skin.

A. Functions of Hair

Schwartz and Rosenblum [35] analyzed the density of body hair on the skin of 21 anthropoid primates and noted that it decreased with increasing size of animals. They also noted that larger animals had lower body surface per body volume. However, larger animals generated considerably more body heat during ordinary physical activity owing to greater muscle mass. Heat of exertion can only be dissipated across the body surface, i.e., through the skin in these animals. In order to maintain cool body temperatures in the absence of sweating, large animals decreased their skin hair density. Alternatively, only those members whose offspring had less body hair survived and progressed through evolutionary time.

Body hair seems to have regressed when human anthropoid primates moved from the forest onto the hunting plains in pursuit of food, where physical exertion generated much muscular heat. The present-day scanty deposits of body hair, however, are not without value. Axillary, torso, and pubic hair evoke strong emotional responses in observers, drawing them into intimate social-sexual contact. And sites where body hair now has receded are important landmarks of skin physiology.

Understanding the distribution, intensity, and pattern of body hair is a way to further assess aging human skin, especially with respect to endocrine variations in male and female status. Estheticians who specialize in electrology and chemical epilation are a valuable resource in this regard. Their professional services when combined with massage increase access to the patron's body surface. Experienced clinicians can evaluate body hair patterns either as a separate procedure, or they can accomplish it during standard professional therapies. Visual patterns alone are sufficient to begin to classify skin, but closer examinations of body hair markedly increase the accuracy and selectivity of skin assessments.

B. Anthropology and Body Hair

Physical anthropologists provide the methods which can be used to classify the patron's skin by visual inspection and hair fiber morphology. An extensive series of studies by Setty provides a valuable description of body hair patterns in male subjects, according to Barker [36].

Clinical observations were made on 13 body sites on 700 white male subjects. The presence of hair on the following sites was noted and scored: *acromial* (top of shoulders), *infracervical* (lower back of neck), *scapular* (shoulder blade region), *lumbar* (loin area), *sacral* (between and just above the buttocks), *infrascapular* (below scapular and above lumbar areas), *circumareolar* (around the nipples, *pectoral* (on the breasts), *sternal* (center or lower part of area occupied by breastbone), *infraclavicular* (below the collarbone), *sagittal* (narrow upward extension of pubic hair along the midline), *acuminate* (triangular upward extension of pubic hair), and *disperse* (distributed over the lower abdomen in a quadrangular pattern). These 13 regions generally account for all hair present on the torso skin of adult males. Among the 700 subjects, substantial numbers were found to have hair at all sites. The region where fewest members had hair was the scapular region (12.3%), whereas 95% had hair on the circumareolar region. All subjects showed hair in *five regions*: dorsal, pubic extension, thoracic, sacral, and abdominal. And five regions of the male body provided the best description of male hair patterns.

Barker's statistical analysis is a prized discovery that can be used by estheticians, skin scientists, biophysicists, and clinicians who work with male patrons. Though female subjects dominate the skin care profession as practitioners and consumers, few references exist to comparable studies done on females.

A general model to categorize human body hair was proposed by Garn [37], who noted unique shapes for the scalp, eyebrows, and eyelashes, beard, body hair, pubic hair, and axillary hair. These sites could be classified as follows: (1) hair not dependent upon steroid hormones, but influenced by changes in growth hormone and sometimes inhibited by androgenic hormones (hair of the head, eyelashes and eyebrows, mid-phalanges, distal segments of the limbs, and lumbosacral triangle; (2) hair dependent upon female steroid hormones (pubic and axillary hair, hair upon the limbs, and hair on the lower abdominal region); and (3) hair dependent upon male steroid hormones (beard, upper lip, nasal tip, and body hair on the back).

These skin regions can be visualized by estheticians and massage therapists, but microscopic examination provides further detailed classification of body hair. Garn [37] states further that hair of the head, eyebrows, and eyelases, pubic area, and axilla do not exhibit enough variation to make them useful in ethnic studies. Hair on the remaining sites does vary significantly with ethnic group, sex, and age. Further examination of 2600 full-length photographs of body build indicated *nine regions* that categorized body hair: *beard* and *upper lip*, *lower abdominal*, *thoracic*, *lower arm* and *leg*, *upper arm* and *leg gluteal*, *lumbosacral*, *lower back*, and *upper back*. These regions exhibited (1) regional independence, (2) racial and individual pattern specificity, and (3) racial and individual variations in density. Phalangeal and ear hairs were added to complete the regional classification of body hair

The pubic hair pattern is markedly different for male and female subjects, according to Dupertuis et al. [38], based upon photographs of 1060 males and 309 females. Four pattern types of pubic hair were noted: *horizontal* (triangle limited to pubic area), *saggital* (thin line of hair ascending from pubic area to umbilicus), *acuminate* (triangle pattern above pubic area peaking at umbilicus), and *disperse* (rectangular pattern covering lower abdominal area). Authors indicated that about 90% of females have the horizontal pattern, with less than 1% showing saggital and disperse patterns. About 10% showed the acuminate pattern. About 20% of the males showed the horizontal pattern and about 50% showed the acuminate pattern. Less than 10% of the males showed the saggital pattern, whereas about 30% of males older than 40 years showed the disperse pattern. No adolescent males showed the disperse pattern.

These reports show that distribution, intensity, and pattern of body hair can be used to classify human subjects. This directs attention to skin because emergence of hair in skin is due to biochemical (steroid) stimulation of target cells, along with a unique predisposition of skin cells to become hair cells in response to stimulation. Categorization of human subjects according to hair patterns, therefore, achieves a classification of skin cells that respond to steroid stimulation.

C. Hair Growth

Influence of aging on hair growth, as reported by Pecararo et al. [39], shows some interesting features that can be used by estheticians to classify their patrons. Axillary hair grows faster in males than in females during years of sexual maturation. Growth rates are lower and similar for preadolescent males and females. The growth rate of hair at the center of the axilla is greater than at the thoracic or brachial subregions for all ages in both sexes. However, the growth rate of central axillary hair decreases in women from the thirteenth week of pregnancy to the fifth week after delivery. Hair density ranges from two to 47 hairs per square centimeter for males and females (no difference), but it decreases steadily with advancing age (10-70 years). The authors suggest that all axillary hairs are not the same; i.e., those of the central region are most sensitive, compared with those of the thoracic and brachial regions.

Rates of hair growth also were studied extensively by Hamilton et al. [40] using chest hairs in males and females. Amounts of hair shaved from a test site on the chest increased with aging between 50 and 70 years. The *number* of hairs peaked earlier at 40-50 years of age. The *percent of gray* chest hair was zero below age 55, but increased steadily thereafter to plateau at about 90 years of age. Gray and pigmented hairs differed in several features, whereas *rate of growth* of chest hair correlated well with that of face (beard) and axillary hair. Also, *length and breadth* of hairs peaked at 40-60 years of age.

The sexual significance of axillary hair was reviewed by Brody [41], who suggests that human sexual behavior is strongly driven by features of the axilla. Secretions of axillary apocrine glands, when acted upon by bacteria, yield odors that become powerful stimuli for social behavior that seeks, or is driven by, a sexual focus. Even though patrons in skin therapy usually present themselves in a hygienic state, or they are provided an opportunity to shower, there is a marked advantage to being able to examine axillary odor. Clinically mature skin therapists can use clinical assessment of axillary odor to note the endocrine and emotional states of patrons. This takes nothing more than knowledge, clinical maturity, opportunity, and acceptance, using one of the most sensitive measuring instruments available, the human nose.

The above review of hair patterns suggests that human skin can be classified (indirectly) by clinically noting the location and intensity of hair on the total body. Estheticians who provide massage services as part of total body skin care can ideally classify skin by noting body hair patterns. Most patrons will fall into broad categories as noted, but some will not. Those unique patrons who do not might be the most interesting and challenging because their skin could be influenced by unusual endocrine states, target cell sensitivity, genetic predisposition, age, and sex, or other physiological factors.

Generally, however, we do not know how to categorize skin in terms of attributes that relate to esthetics and therapy. Yet, there is a clinical feeling that skin types do exist. It is worthwhile for all skin care specialists to become mindful of the total image that their patron's body presents during assessment and treatment. This is the ultimate resource of information for cosmetic chemists, formulators, marketing strategists, and the entire industry.

D. Hair Histomorphology

Body hair is the point of departure for classification of skin (body types) because it has been used for this purpose by numerous physical anthropologists. The simple geometry of scalp hair fibers was used by the anthropologist Hrdy [42] to classify members of seven ethnic groups: *average diameter* was measured by placing a strand of hair between glass plates. A micrometer-equipped microscope was used to scan hair strands from tip to root, and the diameter was measured in microns. *Medullation* was scored at 0 (no medulla), 1 (discontinuous medullation), and 2 (continuous medullation observed on microscopic cross section). *Scale count* was computed as the number of scales per unit length of hair. *Kinking* was defined as sudden constriction and twisting of hair fiber. Each strand was scored 0 (none), or 1 (some). Kinky hair is a commonly used term, but it is a rare feature of human hair. *Average curvature* was estimated by placing hair strands between two glass slides. Curves for distinct regions of the hair strand were drawn on transparent overlays and the radii of these circular curves were measured. Arithmetical mean of all curvatures was computed. *Ratio* of *maximum to minimum curvatures* was computed from the two extreme radii. This index described the regularity of hair strand curl. Regular hair strands will have a ratio of 1; i.e., they will have only one radius of curvature. *Crimp* was defined as the number of times the direction of curvature changed/unit length. This geometrical index is complicated by rotation and lateral displacement of cortices along the hair shaft. And *ratio of natural/straight length* was defined as the projected (short) length/stretched out length. This ratio decreases with increasing curling in which the strand shortens like a coiled spring or wave.

These geometrical attributes of hair can be determined on all body hair, although scalp hair was used to apply these measurements to classify worldwide ethnic groups. Estheticians and cosmetic chemists who wish to specialize in treating skin of the ethnic groups, therefore, can begin to study the patron's and the consumer's skin by using simple geometrical methods to measure body hair. Their findings will provide advanced understanding of human ethnic groups. A special opportunity exists to study general body hair. Like the biophysical studies on human skin, hair studies have been done on only a few body sites.

In this regard, Hrdy showed that world populations consist of far more than just four major groups; i.e., Negroid, Caucasoid, Australoid, and Mongoloid. When hair is used to classify people, it is possible to identify and trace very closely related population groups. This point is illustrated well by Hrdy's study of seven population groups; i.e., Bouganville (Solomon Islands), Malata (Solomon Islands), East African (Kampela, Nairobi), Northwest European (U.S. descendants), Sioux Indians (Peabody Museum hair specimens), Ifugao (Philippines), and Japanese (U.S. descendants/citizens).

Mongoloid hair had large diameter, but African and European hair has about the same diameter. Mongoloid groups also have high indices of medullation, but African and Europeans score low on medullation. Kinking is a notable feature of East African groups. An easily measured property, average curvature, stands out as an important parameter of ethnic hair. East African groups have very high curvature (curly hair), whereas Northwest Europeans, Sioux, and Japanese have straight hair. Ratio of curvature is low for the latter three groups, indicating that their hair is uniformly straight, whereas the other groups exhibit far more diversity in their pattern of hair curl. East Africans are extraordinarily varied in their hair curvature. As one might expect, hair crimp is absent in Japanese, Sioux, and Northwest Europeans, whereas it is present in other groups. The ratio of natural to straight lengths parallels crimp and curvature properties.

This summary of Hrdy's work clearly illustrates the proposal that hair geometry and histomorphology can be used to classify world population groups. Some of these properties are closely related, whereas others are highly divergent. Advanced statistical analyses have been done on these data as part of the physical anthropological tracing of ethnic groups. What one needs now is to study body hair from various sites by these methods and to determine the influence of age, genital sex, endocrine status, and skin treatment. It is likely that hair properties will be sensitive to these factors. Then, it will be possible to classify human panelists, patrons, and consumer groups with regard to their needs for skin treatments and cosmetic products.

IV. PROTOCOL

A. Objectives

1. Identify new skin defects to be rectified.
2. Select agents to treat deficiencies.
3. Specify treatment objectives; define product efficacy in measurable terms.
4. Demonstrate product efficacy; treatment outcomes.
5. Ascertain consumer acceptance; patron satisfaction.

B. Strategy

Studies end, begin, and end with human subjects. Previous study (state of the art) leads to the product to be tested or the skin condition to be treated. The subject's condition either prompts the development of a new product, or a new product concept challenges a persistent skin condition.

C. Plan

1. Functional Age

Select methods from test profiles for pulmonary, respiratory, circulatory, and neural subsystems. Evaluate circadian and menstrual influences on macro- and subsystems.

2. Anthropology

Use body dimensions for size and polymorphism. Estimate androgenic versus estrogenic states. Obtain samples of body hair at various sites. Photograph body and skin sample sites.

3. Skin Type

Use an extensive questionnaire to ascertain sun sensitivity and pigmentation (hair and eyes). Measure reflectance and transmission spectra. Also, measure surface pH and pH reaction, skin bounce, and recovery time of lifted skin-pinch of skin. Compare instrumental data with personal history, self-ratings, and clinical assessments. Make elastomer replicas of the skin surface for topographical analysis. Use adhesive tape to lift off the stratum corneum for particle size analysis.

Treatment plans should contain baselines, positive controls, absolute controls, treatment period, regression, and washout. Double blind and placebo complete this idealized strategy.

Investigators are challenged strenuously by this protocol. The need to learn, however, must be balanced by the reality of costs and the need to satisfy clients. Nevertheless, it cannot be left to other-guy benefactors to conduct research needed for the present-day industry. A practical strategy, therefore, is to "piggy-back" investigations with concurrent testing, product development, subject screening, and patron treatment. This process continuously advances the strategy, leading to accomplishing the above objectives.

REFERENCES

1. Eiermann, H., Consumer surveys to determine moisturizing definition, FDC Reports, The Rose Sheet, December 15, 1986.

2. Borkan, G. A., and Norris, A. H., Assessment of biological age using a profile of physical parameters, J. Gerontol., 35:177-184, 1980.

3. Hollingsworth, J. W., Hashizume, A., and Jablon, S., Correlations between tests of aging in Hiroshima subjects—An attempt to define physiologic age, Y. J. Biol. and Med., 38:11-25, 1965.

4. Furukawa, T., Inoue, M., Kajiya, F., Inada, H., Takasugi, S., Fukui, S., Takeda, H., and Abe, H., Assessment of biological age by multiple regression analysis, J. Gerontol., 30:422-435, 1975.

5. Hochschild, R., The H-Scan—An instrument for the automatic measurement of physiological markers of aging. In Intervention in the Aging Process, part A (W. Regelson and F. Marott Sinex, eds.), Alan R. Liss, New York, 1983.

6. Durnin, J. B. G. A., and Womersley, J., Body fat assessed from total body density and its estimation from skinfold

thickness: Measurements on 481 men and women aged from 16 to 72 years, Br. J. Nutr., 32:77-97, 1974.

7. Brozek, J., and Kinzey, W., Age changes in skinfold compressibility, J. Gerontol., 15:45-51, 1960.

8. Kirk, J. E., and Chieffi, M., Variation with age in elasticity of skin and subcutaneous tissue in human individuals, J. Gerontol., 17:373-380, 1962.

9. Tosti, A., Compagno, G., Fazzini, M. L., and Villardita, S., A ballistometer for the study of the plasto-elastic properties of skin, J. Invest. Dermatol., 69:315-317, 1977.

10. Leveque, J. L., Corcuff, P., de Rigal, J., and Agache, P., In vivo studies of the evolution of physical properties of human skin with age, Int. J. Dermatol., 23:322-329, 1984.

11. Sanders, R., Torsional elasticity of human skin in vivo, Pflugers Arch., 342:255-260, 1973.

12. Grahame, R., and Holt, P. L. J., The influence of aging on the in vivo elasticity of human skin, Gerontologia (Basel), 15:121-139, 1969.

13. Tregear, R. T., Physical Functionsof Skin, Academic Press, New York, 1966.

14. Anderson, R. R., and Parrish, J. A., The optics of human skin, J. Invest. Dermatol., 77:13-19, 1981.

15. Kaidbey, K. H., Poh-Agin, P., Sayre, R. R., and Kligman, A. M., Photoprotection by melanin—A comparison of black and Caucasian skin, J. Am. Acad. Dermatol., 1:249-260, 1979.

16. Zeniske, A., and Krahl, J. A., The occurrence of urocanic acid in sweat, Biochim. Biophys. Acta, 12:479-484, 1953.

17. Leveque, J. L., Poelman, M. C., Legall, F., and de Rigal, J., New experimental approach to measure the skin-reflectance light. Application to cutaneous erythema and blanching, Dermatologica, 170:12-16, 1985.

18. Dawson, J. B., Barker, D. J., Ellis, D. J., Grassam, E., Cotterill, J. A., Fisher, G. W., and Feather, J. W., A theoretical and experimental study of light absorption and scattering by in vivo skin, Phys. Med. Biol., 25:695-709, 1980.

19. Pugliese, P. T., and Milligan, A. J., Ellipsometric measurement of skin refractive index in vivo, Bioeng. Skin. R. Marks, Ed., M.T.P. Press, Lancaster, 1981, pp. 291-302.

20. Solan, J. L., and Laden, K., Factors affecting the penetration of light through stratum corneum, J. Soc. Cosmet. Chem., 28:125-137, 1977.

21. Cook, T. H., Profilometry of skin—A useful tool for the substantiation of cosmetic efficacy, J. Soc. Cosmet. Chem., 31:339-359, 1980.

22. Hoppe, U., Topologie der hautoberflache, J. Soc. Cosmet. Chem., 30:213-239, 1979.

23. Corcuff, P., de Rigal, J., Leveque, J. L., Makki, S., and Agache, P., Skin relief and aging, J. Soc. Cosmet. Chem., 34:177-190, 1983.

24. Van Oort, R. P., Bosch, J. J. T., and Borsboom, P. C., The variation of skin color in different areas of the human body in a caucasian population in CIE 1976 L*, u*, v* color space, J. Soc. Cosmet. Chem., 32:1-14, 1981.

25. Mochizuka, M., Inoue, H., Shaku, M., Koiso, I., Nozawa, S., Kashibuchi, N., Makino, M., Nokano, H., and Kato, S., The physiological condition of skin surface and skin type, Cosmet. Toilet., 97:51-61, 1982.

26. Hartzshtark, A., and Dikstein, S., The performance of a medico-cosmetic consultation unit, Intern. J. Cosmet. Sci., 5:141-151, 1983.

27. Dikstein, S., Instrumental analysis in individual cosmetic consultation, Cosmet. Toilet., 98:77-80, 1983.

28. Potts, R. O., and Buras, E. M., Jr., In vivo changes in the dynamic viscosity of human stratum corneum as a function of age and ambient moisture, J. Soc. Cosmet. Chem., 36:169-176, 1985.

29. Potts, R. O., Buras, E. M., and Chrisman, D. A., Jr., Changes with age in the moisture content of human skin, J. Invest. Dermatol., 82:97-100, 1984.

30. Tagami, H., Electrical measurement of the water content of the skin surface, Cosmet. Toilet., 97:39-45, 1982.

31. Manschot, J. F. M., The mechanical properties of human skin in vivo. Druk: Dissertatiedrukkerij Wibro, Helmond, 1985.

32. Kligman, A. M., Regression method for assessing the efficacy of moisturizers, Cosmet. Toilet., 93:27-35, 1978.

33. Stephenson, L. A., and Kolka, M. A., Menstrual cycle phase and time of day alter reference signal controlling arm blood flow and sweating, Am. J. Physiol., 249:R186-R191, 1985.

34. Turner, C., and Fortney, S., Daily plasma volume changes during the menstrual cycle (abstract), Fed. Proc., 43:718, 1984.

35. Schwartz, C. G., and Rosenblum, L. A., Allometry of primate hair density and the evolution of human hairlessness, Am. J. Phys. Anthropol., 55:9-12, 1981.

36. Barker, D. G., Factor analysis of body hair distribution. Am. J. Phys. Anthropol., 44:27-30, 1967.

37. Garn, G. M., Types and distribution of hair in men, Ann. N.Y. Acad. Sci., 53:498-507, 1951.

38. Dupertuis, C. W., Atkinson, W. B., and Elfman, H., Sex differences in pubic hair distribution, Human Biol., 17:137-142, 1945.

39. Pecoraro, V., Astore, I., and Barman, J. M., Growth rate and hair density of the human axilla, J. Invest. Dermatol., 56:362-365, 1971.

40. Hamilton, J. B., Age, sex and genetic factors in the regulation of hair growth in man. A comparison of Caucasian and Japanese population. In The Biology of Hair Growth (W. Montagna and R. A. Ellis, eds.), Academic Press, New York, 1958, pp. 399-343.

41. Brody, B., The sexual significance of the axillae, Psychiatry, 38:278-289, 1975.

42. Hrdy, D., Quantitative hair form variations in seven populations, Am. J. Phys. Anthropol., 39:7-18, 1973.

Part II

PRODUCT TESTING

7

Noninvasive Methods for Assessing Moisturizers

Gary L. Grove

KGL's Skin Study Center, Broomall, Pennsylvania

I. INTRODUCTION

Although the term moisturizer is a neologism of the cosmetic industry and lacks a precise definition, it is generally agreed that such topically applied products relieve the signs and symptoms of dry skin [1,2]. Despite the fact that dry skin is a troublesome, disquieting problem for many millions of people, surprisingly little is known about its pathogenesis. We do know that dry skin is not a single entity, but rather a family of disorders that can originate in a variety of ways. First, there is a temporary dry skin condition which is the inevitable result of any kind of skin damage whether physical or chemical. This is a stereotypical response to injury and reflects a repair process in which new skin surface cells are formed at a greatly accelerated rate. This accelerated epidermopoiesis causes the skin to be dry and scaly, for example, after sunburn, chemical irritation, abrasion, or detergent damage.

We also know that some of the more severe dry skin conditions are genetically determined. Indeed, the biochemical defect in the rare X-linked ichthyosis has been precisely identified as a lack of a specific enzyme, steroid sulfatase, which is essential for proper utilization of cholesterol in the keratinization process

[3]. Unfortunately, we can only speculate as to what is responsible for more general forms of dry skin that are the primary concern of the typical moisturizer user. For a thorough consideration of the common dry skin problems under discussion in this chapter, one should read the excellent review of Chernosky [4].

Although it is certainly a gross oversimplification, historically, a reduced water content of the stratum corneum has been thought to be the key causative factor in skin dryness. There are two reasons for suggesting this. First consider that in the northeastern United States, dry skin is worse in the winter months when low relative humidity is often accompanied by low temperatures, strong icy winds, and dry, overheated homes (Fig. 7.1). The dry, hot climate of the desert is also apt to provoke dry skin. The critical factor, as shown by the thoughtful analysis provided by Gaul and Underwood [5], is the absolute moisture content of the air. Since the exposed stratum corneum will establish a hydration gradient in equilibrium with the surrounding air, any drop in relative humidity of the environment will lead to a corresponding decrease in the skin surface moisture. As shown by Middleton and Allen [6], the suppleness of the stratum corneum is closely related to its temperature and its water content. This means that such dry skin is relatively inflexible and inelastic under the previously mentioned adverse conditions. As a result, it will crack and fissure in order to accommodate body movements, producing one of the characteristic signs of dry skin.

The second reason for believing that water content is a key factor in dry skin is a series of classic experiments performed by Blank [7,8]. He demonstrated in the early 1950s that skin softening was most effectively achieved with water. Even long-term soaking of cadaver skin samples in various oils failed to produce a comparable degree of softening. Thus, it is not surprising that traditional moisturizers are formulated to increase the moisture content of the stratum corneum in two ways:

1. By occlusion, in which coating the skin with oils such as petrolatum will retard the loss of surface water, thus increasing the moisture content of the underlying stratum corneum.
2. By humectants in which agents, such as glycerin, which draw and strongly bind water, are added to the formulation and thus help trap water in the skin surface.

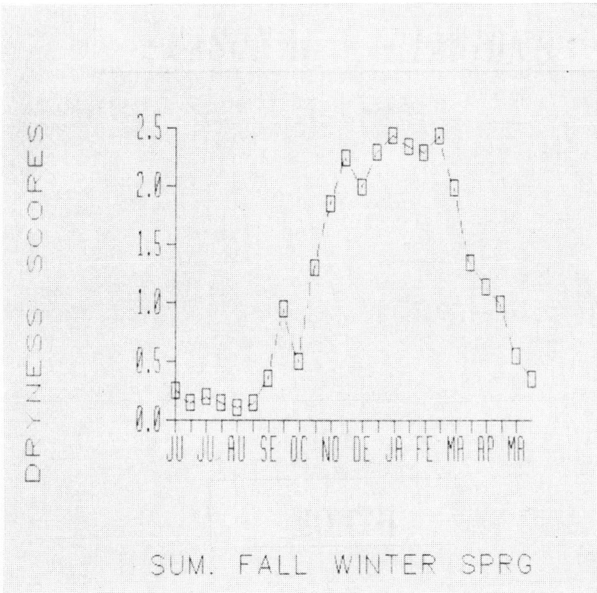

FIGURE 7.1 Seasonal variations of average leg dryness scores.

Of course, one should recognize that water alone can effectively eliminate all visual signs of dryness, although only temporarily. In this context, there are no ineffective cosmetic moisturizers, and the major difficulty facing investigators in this area is not to be unduly influenced by these transient effects. We must also heed the views of Prall et al. [9], who have found very little conclusive evidence to support a causative role of reduced water content with skin flakiness. They feel, and we concur, that the majority of moisturizers work by merely "sticking" loosened squames back onto the surface, thereby altering light scatter, causing the skin surface to appear more transparent, and hence less dry for a short period of time. We [1,2] classify such products as being "cosmetic" moisturizers. On the other hand, we have been able to identify "therapeutic" moisturizers which in fact modify the dry skin process rather than simply conceal it. For example, topically applied alpha-hydroxy acids are known to be effective in alleviating the signs and symptoms of dry skin in ichthyosis [10]. How this is achieved is not completely understood, but it is unlikely that either occlusion or humectancy are involved.

With these points in mind, let us review some of the clinical and instrumental methods that have been developed to quantify skin dryness, and thus allow us to evaluate moisturizers.

II. THE REGRESSION METHOD

By far the most widely used clinical method for assessing the efficacy of moisturizers is the regression method first proposed by Kligman [1,2]. We have had considerable experience with this method, and have modified certain aspects to further enhance its usefulness in identifying products that are therapeutic moisturizers which really modify the dry skin process, as opposed to cosmetic moisturizers which merely conceal it temporarily. The most noteworthy change is our preference for middle-aged suburban housewives over young college co-eds as panelists. Although dry skin occurs across all age groups, we have found the housewives to have more consistent levels of dryness, and most importantly, to be far more reliable in complying with the requirements of the protocol, which includes refraining from the use of any products except those provided during the study. We have also found that more mature volunteers are quite capable of self-administering the treatments whether using a controlled dosage or ad libitum application of the products. Nevertheless, we still prefer to have panelists report to our research center on a daily basis for at least their AM applications.

In recruiting volunteers for these types of clinical studies, we screen for those individuals who have a history of dry skin problems of sufficient severity to require routine use of moisturizers during the dry skin season. We have found it best to enroll approximately twice as many subjects as needed, and require them to refrain from the use of all products on the test sites for a 2-week pretrial period. Since many of our studies are run on the lateral aspects of the leg, we also impose restrictions on the scheduling and manner in which leg hair is removed. Restrictions are also placed on the use of such products as Buf-Pufs® (3M, St. Paul), loofa sponges, and bath oil beads.

At the end of this 2-week pretrial period, we select from these candidates the best subjects who have appreciable signs of dryness of equivalent magnitude on the paired test sites. This pretrial period also gives us a chance to determine the reliability of the panelists and their willingness to fully comply with the requirements of the experimental design, especially the restrictions.

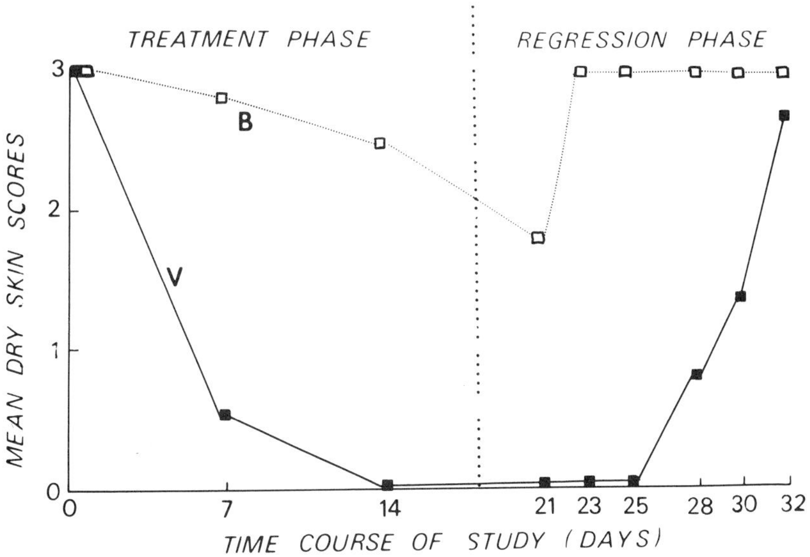

FIGURE 7.2 Comparative behavior of a "therapeutic" (product V) and a "cosmetic" (product B) moisturizer as revealed by the Modified Kligman Leg Regression Test.

Once a panel of 20 suitable subjects is selected, they are treated twice daily with the appropriate test products on each weekday for the next 2 weeks. Both applications are made by a technician, who applies liberal amounts of product with moderate hand pressure for a period of 30 sec per leg. Plastic gloves are worn to prevent cross contamination of test products. Of course, in-lab studies with controlled dosing or at-home studies with ad libitum applications by the panelists themselves could also be conducted.

The test sites are evaluated on the first two Mondays (days 7 and 14), and the study is terminated if improvement is negligible on the second Monday. Otherwise, follow-up observations are made on a daily basis every weekday for the next 2 weeks, or until the original level of dryness appears. During this regression period, the application of test products is discontinued, but all restrictions as to how the test sites are to be treated are maintained. Typical results based on the classic 6-week Kligman schedule which illustrates the behavior of a poor moisturizer and a very effective moisturizer are shown in Fig. 7.2. It is also important to note that the Unilever group [9] has

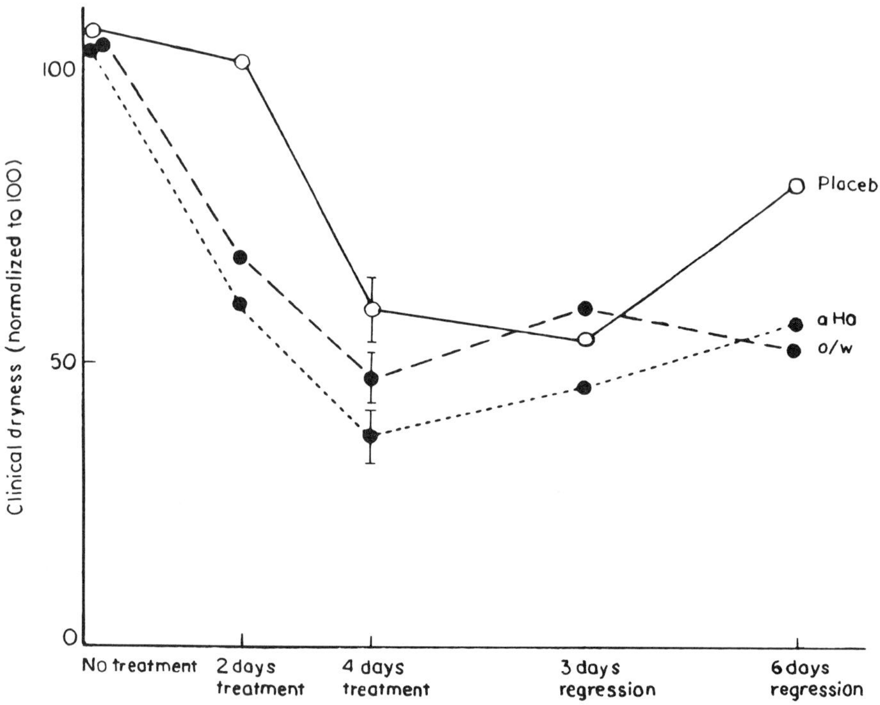

FIGURE 7.3 Comparison of three formulations using the short-term variant of the Regression Test as developed by J. K. Prall of Unilever Research. (From Ref. 9.)

created a short-term variant that compressed the treatment and regression phases into a single week. As can be seen from Fig. 7.3, this works quite well for pilot studies, but we still feel it is best to run the more extended study for claim substantiation purposes.

This basic design, whether short term or extended, can be followed regardless of whether the test sites are the legs, arms, elbows, faces, or other sites. The clinical expression of dry skin does vary from site to site, and the grading scale should be adjusted accordingly to reflect these differences. One effective way to do this is to utilize a comparative rating scheme in which the grader is forced to choose which site is better and indicate by how much better using the terms slight, moderate, or dramatic to describe the degree of difference. Another equally valid approach is to use a well-characterized scoring scheme that has a sufficient number of grades (6 or more) to allow good resolution between products. We are especially impressed with the approach developed by Spencer's group [11] in which a reference set of standardized photos is used to visually define the grading scheme for the various attributes of dry skin being evaluated. Not only do such photographs serve as references for the expert graders to utilize during their actual assessments, but they can also serve in training and rating their proficiency through an external review process. Moreover, with such a photographic reference system, a much clearer understanding of the inclusion/exclusion criteria can be established.

It is also important to note that the relative merits of the test formulations as perceived by the panelists should also be ascertained. This can be done by a forced choice procedure, comparative visual analog scales, or written questionnaire.

III. INSTRUMENTAL METHODS

Because of the increased demand for scientific documentation of advertising claims made for skin care products such as moisturizers, there has been considerable interest in developing instrumental methods for quantifying product efficacy in vivo [12-14]. The use of instrumental testing techniques by a number of premier companies has provided them with a significant marketing edge by providing functional claims and supportive research data to the consumers. As a result, we have seen a large number of reports extolling the virtues of a wide variety of techniques which purportedly can be used to evaluate skin care product performance. In the following sections we will briefly review the most promising of the methods for their utility in documenting the efficacy of moisturizers. The emphasis will be on brevity, and the reader will be referred to appropriate reviews and research papers for details.

A. Profilometry of Skin Surface Replicas

There are a number of different ways by which the skin surface topographic features can be noninvasively analyzed for degree of roughness. The reader is referred to a comprehensive review [15] for details of the present state of the art.

One of the most widely used methods has been to cast a replica of the skin surface by using one of the many excellent silicone rubber impression materials such as Silflo. A plastic positive can then be cast and measured with a computerized stylus instrument that provides a contour tracing of the surface. A number of different data formats are possible with such instruments, ranging from a simple plot of contour along a single axis to a three dimensional stereograph of the skin surface topography. Provided that the movement of the stylus is sufficiently accurate, the problem becomes which of the many parameters that can be derived from such specimens are the best measure of skin surface roughness. As pointed out by both Cook [16,17] and Makki [18], certain roughness parameters appear to be more utilizable than others, and can vary with scanning orientation and body area tested.

Despite its popularity, we do not advocate the use of replica profilometry to evaluate moisturizers, especially if the panelists have flakes and scales. There are many reasons for feeling this way. Among them is the fact that the application of the silicone rubber material most surely flattens and disturbs the uplifted scales.

B. Scanning Microdensitometry of Macrophotographs

Several investigators [11,19] have shown that the visual benefits provided by a moisturizing cream can be photodocumented. Several attempts have been made by Marshall's group [20,21] to provide a more quantitative assessment of skin surface texture from such low magnification photographs. This includes using a scanning microdensitometer to detect the shadows and highlights of the skin surface captured in the photographic negative taken under standardized lighting conditions. The resulting contour line is quite similar to those obtained in replica profilometry, and calculation of roughness parameters can be done in much the same way. Although scanning densitometry has apparently been effective in following the clinical progression of patients with scaling disorders, such as psoriasis or ichthyosis, it seems less able to discriminate the degree of surface roughness in normal volunteers before and after application of various occlusive emollient agents. Thus, it seems that additional improvements must be made if this technique is to acquire the resolution needed to deal with more typical dry skin problems.

C. Squametry of Tape Strippings

The use of adhesive tape strippings to facilitate observations of the skin surface was first reported by Wolf [22]. When tape is pressed against the skin, the outermost, loosely adherent portion of the stratum corneum will stick to the tacky adhesive. Thus, upon removal the tape provides a specimen which retains the topographical relationships of the skin surface and its pattern of desquamation (Fig. 7.4). This technique has been called squametry by Prall [9], and has been used extensively by both his group and ours [23] to provide an index of skin surface scaliness. Although there are a few subtle differences in our approaches, the basic strategy is still the same. Both of us use an image-processing procedure to enhance and segment out the uplifted scales captured in such specimens. These objects of interest can be sized and counted to give a measure of the degree of skin surface scaliness.

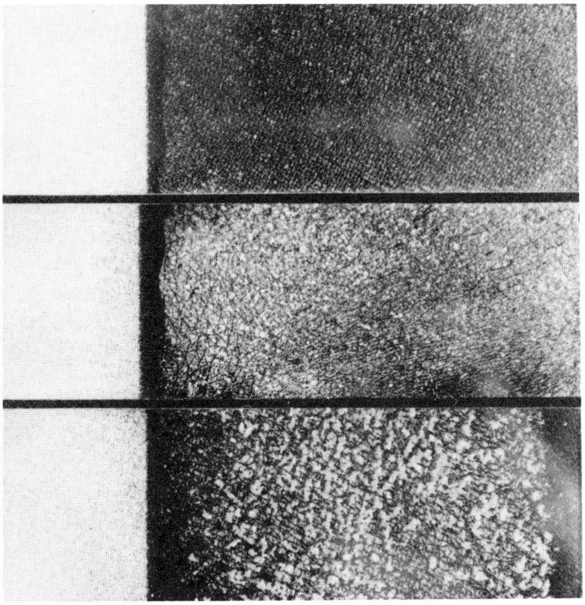

FIGURE 7.4 Adhesive tape strippings removed from the lower legs of a dry skinned and normal individual.

D. In Vivo Image Analysis

Since an image-processing system has a video camera input, it is possible to directly record skin surface features. Indeed, we have been able to do a near real-time analysis of skin surface roughness with our Magiscan system [23]. To do so in a reliable fashion requires extreme attention to standardized lighting and critical camera angles. The use of special filters, such as those suggested by Dent [24], can also be used to enhance surface detail. Unfortunately, this all takes time, and we have found it more productive to take an intermediate photograph with a conventional 35-mm S.L.R. camera system, similar to that described by Prall's group [9]. This means that this approach is converging on the macrophotographic technique described above. We feel very optimistic that our application of the digital image-processing technique to enhance and analyze these types of specimens will make this a more useful technique to evaluate low to moderate levels of dryness than it has been heretofore.

E. Coefficient of Friction Devices

Another instrumental approach that has received considerable attention as a possible method for objectively evaluating skin roughness involves measuring the coefficient of friction [25]. Devices which have been employed for this purpose include a rotating wheel [26,27], a revolving ground glass disc [28,29], a sliding sled [30], and a modified viscometer [31,32]. In all cases, the underlying principle is that frictional properties of human skin in vivo can be assessed by determining how much force is required to drag an object across the skin surface. Although Weinstein [30] states rather matter of factly that "smoother skin requires less force," in actual practice the interpretation of differences in frictional properties induced by product application is very complicated. For example, moisturizers which smooth and hydrate the skin can actually increase friction as a result of increased contact area with the moving surface of the measuring device [26,28,29,33,34]. Nacht and coworkers [35] have suggested that these changes might also reflect an increased adhesiveness of the stratum corneum in a hydrated state. On the other hand, materials which act as surface lubricants do indeed lower the coefficient of friction, making the skin feel more slippery [26,34,36,37]. In the case of a greasy occlusive material such as petrolatum, a biphasic response can be observed in which initially the coefficient of friction decreases owing to its lubricant properties, whereas later on it increases over baseline as a consequence of the increased hydration induced by occlusion [34]. Thus, it is not surprising that coefficient of friction measurements by themselves sometimes correlate very poorly with sensory scores of smoothness as shown by Prall's group [28]. While additional studies in this area may correct this situation, at the present time, such measurements do not seem to be useful in providing a precise and objective measure of skin roughness per se. Nevertheless, the use of this approach to screen topicals for their after-feel attributes of greasiness and their moisturizing properties as outlined by Nacht and coworkers [35] should prove worthwhile.

F. Scratch Resistance Test

Prall [28] has demonstrated that one of the physical properties of the skin which contributes to the overall perception of smoothness by the customer is hardness, as measured by the lowest pressure load and which causes a stylus to just visibly scratch the skin surface. Although not widely employed, this relationship forms the basis of a recent ad campaign that has proven highly successful. Indeed, many individuals who suffer from dry skin problems will do an analogous procedure by scraping the skin surface with their fingernail to reveal underlying defects that might not be so obvious at first glance.

G. Sonic Wave Propagation

Experimental results from a number of laboratories have indicated that the mechanical properties of the skin vary dramatically with the stratum corneum hydration. These are often associated with subjective assessments of moisturization, especially with regard to the ability of the product to make the skin feel "soft." For example, Torgalkar [38] utilized a vibrational device operating in the audible frequency range (approximately 700 Hz) to impart small amplitude oscillations normal to the skin surface. By scanning the frequency range, he was able to measure the resonant frequency of these "ripple waves" and from the known vibrational characteristics of the oscillator device calculate the energy loss of the viscous component of the skin. His results revealed that a continuous and rapid decline in energy loss occurred immediately after removal of an occlusive wrap that remained in place on the volar forearm of a subject for 14 hr. Indeed, a constant value was reached within 10 min

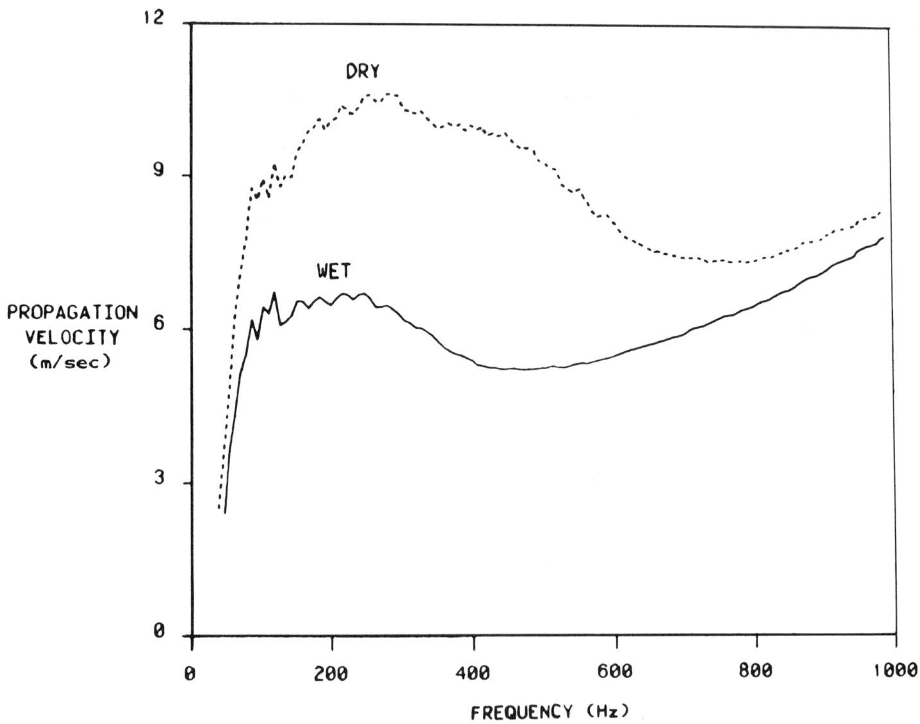

FIGURE 7.5 The propagation velocity versus frequency for shear waves in the skin of the back of the hand for one individual. Data were obtained under ambient conditions (DRY) or after soaking the hand in water for 5 min followed by brief toweling (WET). (Original illustration was prepared by Potts; from Ref. 40, used with permission.)

after the wrap was removed, suggesting that the outer epidermal layers were primarily responsible for these changes in "softness."

More recently, Potts and his coworkers [39,40] have extended these studies of the effects of moisturizers on the viscoelastic properties of the stratum corneum. Their device consists of a vibrational stylus that lightly rests on the skin surface. The amplitude of the vibrations propagated on the skin surface is constant over the broad range of frequencies (20-1000 Hz) utilized in this method. A second pick-up stylus is positioned on the skin surface a few millimeters from the first. By using a spectrum analyzer, it was possible to determine the time required for these shear waves to travel through the skin surface and the degree of amplitude dampening as a function of frequency. As demonstrated in Fig. 7.5, the propagation velocity of hydrated skin was dramatically reduced, especially at the lower frequencies where the properties of the stratum corneum are primarily measured. This group [41] has also shown there to be seasonal and age-related changes that are consistent with the notion that this approach is indeed an indirect measure of the hydration state of the outer layers of the stratum corneum, although the exact relationship between frequency and depth needs to be established.

Dahlgren and Elsnau [42] have utilized a similar sonic velocity technique to assess the relative efficacy of moisturizers. Their results have shown that topically applied moisturizers do indeed decrease the sonic propagation velocity in skin and that this decrease is highly correlated with subjective assessments of moisturization. More importantly, the instrumental changes were noted after only a few days of use, whereas the subjective assessments took several weeks, indicating that sonic velocity can be a predictive measure.

H. Gas-Bearing Electrodynamometer

The gas-bearing electrodynamometer (GBE) developed by Hargens [43] has proven to be quite useful in determining the viscoelastic properties of skin. The key element of this device is a highly compliant, virtually frictionless bearing of pressurized gas. Suspended upon this "gas bearing" is an armature which mounts both the force coil and the core of a linear displacement transducer. The arrangement is such that the GBE measures the displacement of skin in response to a rapidly oscillating force placed parallel to its surface. The resulting dynamic stress-strain diagram (hystersis loop) which instantaneously appears on an oscilloscope can be analyzed to reveal several characteristics such as stiffness, softness, and compliance. It is important to note that the GBE does not measure a fundamental property of the stratum corneum because there is a dermal component to these measured properties. Nevertheless, both Christensen et al. [44] and Cooper et al. [45] have observed a dramatic decrease in the elastic modulus immediately after the application of water to the skin. Such a rapid response could not result from the action of the applied water on the dermis

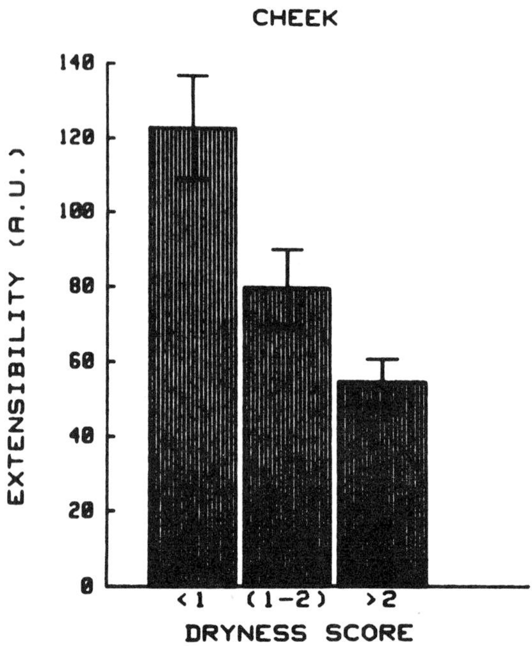

FIGURE 7.6 Relationship between skin extensibility as measured by the narrow gap twistometer and facial dryness scores. (From Ref. 48, used with permission.)

and clearly indicates that the mechanical properties of the stratum corneum contribute significantly to the elastic modulus. Indeed, additional studies by both groups [44,45] have shown there to be a high degree of correlation of elastic modulus measurements with visual assessments of skin condition by a trained grader. In vitro studies [46] have also shown that dry skin is generally stiffer than normal skin, and effective treatments such as with glycerin can indeed soften the skin.

I. Twistometre

The group of Leveque and de Rigal [47] have made extensive use of a torsional technique to measure in vivo the influence of stratum corneum hydration on its extensibility. Their device, which is named the Twistometre® (L'Oreal, Paris), applies a weak torque to a rotating disc that is attached to the skin with a nonslip, tacky adhesive. The torque is held constant and the movement of the disc monitored by a rotational sensor that is linked to a microprocesser that computes the main parameters. The area subjected to this twisting load is well delineated by a fixed guard ring that is concentric to the inner rotating disc. When the distance between the disc and surrounding guard ring is less than 1 mm, the extensibility measurement (Ue) primarily reflects the resistance of the stratum corneum to stretching. Both short- and long-term studies with a variety of moisturizers and humectants have shown that this is a sensitive measure for rating such products as to their hydrating efficacy [47]. It has also been very clearly shown (Fig. 7.6) that there is a striking decrease in stratum corneum extensibility as the severity of dryness increases [48].

J. Electrical Properties

As pointed out in the excellent review by Leveque and de Rigal [49], the flow of electrical current through the skin surface is related to the water content of the stratum corneum, and thus offers a noninvasive method for assessing moisturization. They go on to describe that there are three distinct ways in which the application of an electrical field to the stratum corneum results in current flow. They are:

1. Orientation of dipole moments of various constituents such as keratin
2. Ionic movement within the stratum corneum
3. Water mobility and proton exchange within the stratum corneum

It is obvious that water directly influences only the third mechanism, but it also indirectly facilitates current flow by enhancing dipole motion and ion mobility owing to decreased viscosity in hydrated stratum corneum. Unfortunately, agents other than water can lower skin impedance or resistance to flow. For example, urea, a common component in moisturizers, can induce changes in keratin dipole orientation by virtue of its protein denaturant properties. In addition, salts, whether as components of moisturizers or derived from perspiration, are intrinsically mobile and will cause a dramatic decrease in stratum corneum impedance.

Despite these problems, a number of investigators have found electrical measurements to be useful in assessing skin moisturization. Among the early workers was Clar [50], who

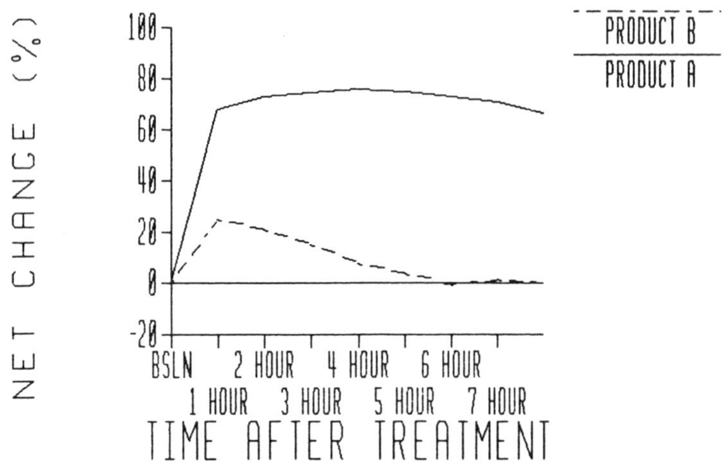

FIGURE 7.7 Impedance and stratum corneum hydration changes with time following application of moisturizers A and B.

worked at low frequencies where the net impedance of the stratum corneum is quite high. Unfortunately, her device required liquid junction electrodes to assure adequate current flow. Not only do these wet electrodes directly affect the hydration state of the underlying skin, they are also occlusive and must remain in place for approximately 20 min for each determination.

More recently, investigators have employed higher frequency impedance measurements that allow the use of dry electrodes. The most notable of these devices is the unit developed by Tagami and coworkers [51-53] that is now commercially available as the Skicon-200. This instrument operates at 3.5 MHz and uses an electrode of two concentric brass cylinders separated by a phenolic insulator. Their results from both in vitro and in vivo experiments show that impedance drops with increasing hydration. Furthermore, this method has been used by Tagami's group [53] and ours [54] to evaluate moisturizers as to their relative efficacy and duration of effect. For example, Fig. 7.7 shows such a time course in which it is clear that moisturizer A clearly outperforms moisturizer B in both regards.

Other devices based on electrical measurements such as the Nova Dermal Phase Meter [55,56] and the laboratory-constructed wire mesh electrode instrument of Serban's group [57,58] have also been used to assess skin dryness and the effects of chronic

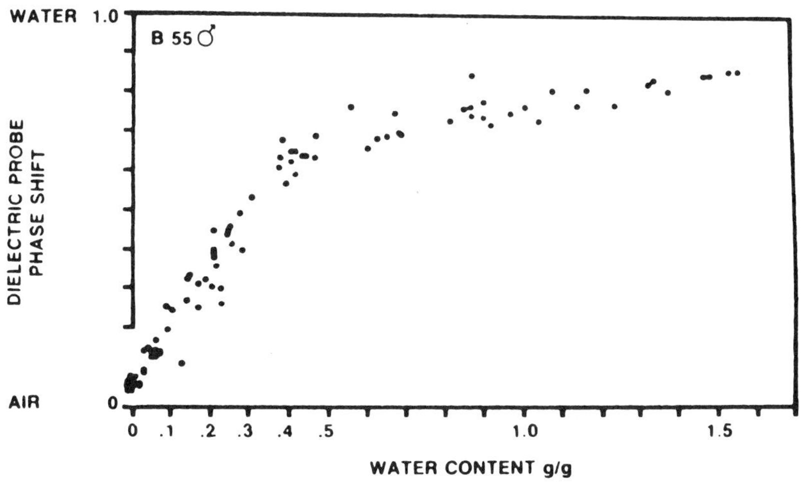

FIGURE 7.8 In vitro calibration of the focused microwave probe response. The probe response is expressed in lab units, where air yields 0 and contact with water yields 1. The water content was determined gravimetrically for SC samples obtained by trypsin treatment of skin from a 55-year-old black male. Original figure was prepared by Potts using data derived from the Ph.D. thesis of S. L. Jacques. (From Ref. 40, used with permission.)

treatments with such agents as creams, lotions, and raw materials. Although it is assumed that these measurements are of the most superficial layers of the skin surface, the depth of the stratum corneum being probed with any of these electrical devices has not been well established. Jacques and his coworkers [59-61] have begun to address this problem by utilizing a very high-frequency microwave device to measure the dielectric properties of the skin. They are operating in the region where the electrical fields will cause the water molecules to oscillate, and the amount of energy required to do so is dependent upon their number and hence their concentration. Clearly, microwaves can deeply penetrate within tissue; indeed, this is the basis for microwave ovens. To confine the microwave field to within a shallow depth of just a few microns, a focused microwave probe based upon fringe fields between closely spaced electrodes was created. The ability of the microwave probe to measure differences in skin surface water content was documented in vitro by following the water uptake of samples of human stratum corneum

(Fig. 7.8). In good agreement with other findings, these data showed biphasic behavior in which water was tightly bound at hydration levels of less than 30% (w/w), whereas it behaved like bulk liquid at higher levels. More recently, the focused microwave probe was employed to measure the water concentration profile of human skin in vivo. This stratagraphic analysis was achieved by spacing the probe away from the skin surface with sections of inert Teflon film of varying thickness. As anticipated, and consistent with Fick's law, a nonlinear water concentration gradient across the stratum corneum was revealed in this manner.

K. Photoacoustic Spectroscopy

Another technique that offers the potential for quantitatively measuring the water concentration profile is photoacoustic spectroscopy (PAS). For a review of the clinical applications of this PAS method, the reader is referred to several excellent reviews [62,63,64]. Briefly, PAS measures the thermal properties of the stratum corneum by modulating optical radiation with a beam chopper to create periodic heat waves within this compartment. These heat waves propagate through the stratum corneum, giving rise to periodic pressure waves that can be detected at the skin surface by an extremely sensitive microphone. The distance a heat wave can travel before being dissipated is related directly to the thermal diffusity of the tissue and inversely to the modulation frequency. This means that by altering the modulation frequency of the optical radiation, one can probe the thermal properties of the stratum corneum as a function of depth.

Although such measurements are experimentally difficult and time consuming, the effects of hydration-induced changes in the human stratum corneum water gradient has been evaluated both in vitro [65] and in vivo [66,67]. The results revealed that the stratum corneum is nonhomogeneous, and that the outermost surface layers dehydrate more rapidly than underlying tissue.

L. Infrared Spectroscopy

Attenuated total reflection infrared spectroscopy has been used by a number of investigators [68-71] to noninvasively examine the physicochemical interactions of skin and topically applied products. The basic premist of this approach is that the infrared (IR) absorbence spectrum of water can be, at least in principle, uniquely identified and effectively separated from the IR absorbence spectrum of everything else in the skin plus topical product system. Once this is achieved, the water content can be quantitatively determined from the intensity of IR absorbence in that particular region.

In early applications of this technique, the O-H stretching band from 3200 to 3600 cm was utilized to assess skin hydration [72]. More recent workers [73-75] have used the ratio of the amide I band at 1645 cm to the amide II band at 1545 cm to evaluate moisturizer efficacy. The idea here is that the amide I band is due to both protein and water absorbencies, whereas the amide II band is due to protein alone. Unfortunately, as pointed out by Potts [76], it seems that both the amide I and II bands of keratin, the major protein component of the stratum corneum (SC), change with hydration, thus removing the possibility of an absolute determination. Moreover, many moisturizers contain absorbencies which overlap with and therefore influence the intensity of the amide I band. Thus there are a number of difficulties inherent in quantitative evaluation of stratum corneum water content from amide I/II intensity ratios.

Potts [77] is a strong advocate of the use of spectroscopy, and feels that these problems can be overcome by using a ZWS crystal that has a low refractive index which closely approximates that of skin and enhances sensitivity. His Fourier Transform IR (FTIR) spectrophotometer allows rapid data collection and analysis. Moreover, he has chosen to use an incidence angle close to the critical angle to enhance the depth of penetration of the iodination into the skin and an optical path enclosed in a dry nitrogen purge to eliminate absorbencies due to atmospheric water vapor. With this arrangement, a minor absorbence band of water centered at 2100 cm has been found most useful in determining a quantity relative to the hydration of the stratum corneum.

M. Evaporimeter Measurements

The development of various methods for measuring transepidermal water loss (TEWL) have been comprehensively reviewed [78-81]. One would expect that the greater the occlusivity of the product, the greater the reduction in TEWL will be upon its application, and this is certainly the case as shown by several workers [82-86]. With the commercially available Servo-Med Evaporimeter, these types of assessments are straightforward and easy to accomplish. Some care should be taken to ensure that the amount of product applied is relevant to the intended use conditions, as a thick film can give an unrealistically high value [87], and that the ambient temperatures are low so that sweating is not a factor [88].

In dealing with moisturizers that are based on humectants, the situation is much more complex. Very early studies by Powers and Fox [89] demonstrated that transepidermal water loss was increased not decreased when treated with effective moisturizers. Rietachel [90,91] has confirmed these findings, especially with regard to the behavior of products containing glycerin. It is known that the diffusion coefficient increases with increased hydration of the stratum corneum [92,93], and thus water vapor will move through more readily. This means that more water is available to the outermost layers of the stratum corneum, and thus leads to relief of the signs and symptoms of dry skin. Unfortunately, damaging products such as detergents will disrupt the barrier properties of the stratum corneum and also lead to increased flux and elevations in transepidermal water loss values. Of course, this is an undesirable consequence of product usage.

IV. SUMMARY

The methods for assessing the moisturizing efficacy of skin care products range from relying on the subjective assessments made by the panelists of their own skin to more objective computer-assisted instrumental methods. No matter how accurate and precise an objective measurement may be, the question of relevance must always be considered. The measurement must ultimately correlate with the clinical perception of dry skin. At the moment, no one instrumental method can replace the experienced grader in rating dry skin. Devices invariably measure a single attribute, whereas the human brain integrates multiple inputs secured from vision and touch. By employing a battery of instrumental methods we can hope to gain a much fuller understanding of what defects are responsible for the dry skin problem, but this should be done in conjunction with, not instead of, the more classic clinical studies. Indeed, we strongly advocate that a three-pronged approach based on panelist self-appraisal, expert grader evaluations, and relevant instrumental measurements be utilized for adequately substantiating product performance.

REFERENCES

1. Kligman, A. M., Regression method for assessing the efficacy of moisturizers, Cosmet. Toilet., 93:27-35, 1978.

2. Kligman, A. M., Lavker, R. M., Grove, G. L., and Stoudemayer, Some aspects of dry skin and its treatment. In Safety and Efficacy of Topical Drugs and Cosmetics (A. M. Kligman and J. J. Leyden, eds.), Grune and Stratton, 1982, pp. 221-238.

3. Shapiro, I. J., Weiss, R., and Webster, D., X-linked ichthyoses due to sterol sulfatase deficiency, Lancet, 1:70-72, 1970.

4. Chernosky, M. E., Clinical aspects of dry skin, J. Soc. Cosmet. Chem., 27:365-376, 1976.

5. Gaul, L. E., and Underwood, G. B., Relation of dewpoint and barometric pressure to chapping of skin, J. Invest. Dermatol., 18:9-12, 1951.

6. Middleton, J. D., and Allen, B. M., The influence of temperature and humidity on stratum corneum and its relation to skin chapping, J. Soc. Cosmet. Chem., 24:239-243, 1973.

7. Blank, I. H., Factors which influence the water content of the stratum corneum, J. Invest. Dermatol., 18:433-440, 1952.

8. Blank, I. H., Further observations on factors which influence the water content of the stratum corneum, J. Invest. Dermatol., 21:259-271, 1953.

9. Prall, J. K., Theiler, R. F., Bowser, P. A., and Walsh, M., The effective of cosmetic products in alleviating a range of skin dryness conditions as determined by clinical and instrumental techniques, Int. J. Cosmet. Sci., 8:159-174, 1986.

10. Van Scott, E. J., and Yu, R. J., Control of keratinization with alpha-hydroxy compounds, Arch. Dermatol., 110:586-590, 1974.

11. Seitz, J., Rizer, R., and Spencer, T., Photographic standardization of dry skin, J. Soc. Cosmet. Chem., 35:423, 1984.

12. Grove, G. L., Techniques for substantiating skin care product claims. In Safety and Efficacy of Topical Drugs and Cosmetics (A. M. Kligman and J. J. Leyden, eds.), Grune and Stratton, 1982, pp. 157-176.

13. Leveque, J. L., Physical methods for skin investigation, Int. J. Dermatol., 22:368-375, 1983.

14. Grove, G. L., Design of studies to measure skin care product performance, Bioeng. Skin, 3:359-373, 1987.

15. Grove, G. L., and Grove, M. J., Objective methods for assessing skin surface topography noninvasively. In Cutaneous Investigation in Health and Disease (J. L. Leveque, ed.,), Marcel Dekker, New York, 1988, pp. 1-32.

16. Cook, T. H., Profilometry of skin: A useful tool for the substantiation of cosmetic efficacy, J. Soc. Cosmet. Chem., 31:339-359, 1980.

17. Cook, T. H., Craft, T. J., Brunelle, R. L., Norris, F., and Griffin, W. A., Quantification of the skin's topography by skin profilometry, Int. J. Cosmet. Sci., 4:195-205, 1982.

18. Makki, S., Agache, P., Mignot, J., and Zahouani, H., Statistical analysis and three dimensional representation of the human skin surface, J. Soc. Cosmet. Chem., 35: 311-325, 1984.

19. Quattrone, A., and Laden, K., Physical techniques for assessing skin moisturization, J. Soc. Cosmet. Chem., 27:607-623, 1976.

20. Marshall, R., and Marks, R., Assessment of skin surface by scanning densitometry of macrophotography, Clin. Exp. Dermatol., 8:121-127, 1983.

21. Marshall, R., and Marks, R., Quantification of skin surface texture of macrophotography and computer aided scanning densitometry, J. Audiov. Media Med., 6:98-103, 1983.

22. Wolf, J., Das oberflachen relief der menschlichen Haut, Z. Mikrosk. Anat. Forsch, 47:351, 1940.

23. Grove, G. L., Dermatological applications of the Magiscan image analysing computer. In Bioengineering and the Skin (R. Marks and P. A. Payne, eds.), MTP Press, Lancaster, England, 1981, pp. 173-182.

24. Dent, R. V., The photographic aspects, J. Lab. Clin. Med., 26:1852-1862, 1983.

25. Wolfram, L. J., Friction of skin, J. Soc. Cosmet. Chem., 34:465-476, 1983.

26. Comaish, S., and Bottoms, E., The skin and friction: Deviation from Amonton's laws, and the effects of hydration and lubrication, Br. J. Dermatol., 84:37, 1971.

27. Comaish, S., Harborow, P. R. H., and Hofman, D. A., A hand-held friction meter, Br. J. Dermatol., 89:33, 1973.

28. Prall, J. K., Instrumental evaluation of the effects of cosmetic products on skin surfaces with particular reference to smoothness, J. Soc. Cosmet. Chem., 24:693-707, 1973.

29. Highley, D. R., Coomey, M., DenBeste, M., and Wolfram, L. J., Frictional properties of skin, J. Invest. Dermatol., 69:303, 1977.

30. Weinstein, S., New methods for the in-vivo assessment of skin smoothness and skin softness, J. Soc. Cosmet. Chem., 29:99-115, 1978.

31. Gerrard, E. D., and Stimpson, I. M., A versatile friction meter based on a viscometer, Lab. Pract., 33:82-83, 1984.

32. Gerrard, E. D., and Stimpson, I. M., The effect of treatment on skin friction coefficient in vivo, Bioeng. Skin, 1:229, 1985.

33. Naylor, P. F. D., The skin surface and friction, Br. J. Dermatol., 67:239, 1955.

34. El-Shimi, A. F., In vivo skin friction measurements, J. Soc. Cosmet. Chem., 28:37, 1977.

35. Nacht, S., Close, J., Yeung, D., and Gans, E. H., Skin friction coefficient: Changes induced by skin hydration and emollient application and correlation with perceived skin feel, J. Soc. Cosmet. Chem., 32:55, 1981.

36. Cussler, E. L., Zlotnick, S. J., and Shaw, M. C., Texture perceived with fingers, Percept. Psychophys., 21: 504, 1977.

37. Appeldoorn, J. K., and Barnett, G., Frictional aspects of emollience, Proc. Sci. Sect. Toilet Goods Assoc., 40:28, 1963.

38. Torgalkar, A. M., A resonance frequency technique to determine the energy absorbed in stratum corneum, in vivo. In Bioengineering and the Skin (R. Marks and P. A. Payne, eds.), MTP Press, Lancaster, England, 1981, pp. 55-65.

39. Potts, R. O., In vivo measurement of water content of the stratum corneum using infrared spectroscopy: A review, Cosmet. Toilet., 100(10):27-31, 1985.

40. Potts, R. O., Stratum corneum hydration: Experimental techniques and interpretations of results, J. Soc. Cosmet. Chem., 37:9-33, 1988.

41. Potts, R. O., Buras, E. M., and Chrisman, D. A., Changes with age in the moisture content, J. Invest. Dermatol., 82: 97-100, 1984.

42. Dahlgren, R. M., and Elsnau, W. H., Measurement of skin condition by sonic velocity, J. Soc. Cosmet. Chem., 35:1-20, 1984.

43. Hargens, C. W., The gas-bearing electrodynamometer (GBE) applied to measuring mechanical changes in skin and other tissues. In Bioengineering and the Skin (R. Marks and P. Payne, eds.), MTP Press, Lancaster, England, 1981, pp. 113-122.

44. Christensen, M. S., Hargens, C. W., Nacht, S., and Gans, E. H., Viscoelastic properties of intact human skin: Instrumentation, hydration effects, and the contribution of the stratum corneum, J. Invest. Dermatol., 69:282-286, 1977.

45. Cooper, E. R., Missel, P. J., Hannon, D. P., and Albright, G. B., Mechanical properties of dry, normal, and glycerol-treated skin as measured by the gas-bearing electrodynamometer, J. Soc. Cosmet. Chem., 36:335-348, 1985.

46. Missel, P. J., Bowman, W. D., Benzinger, M. J., and Albright, G. B., An in vitro method for skin preservation to study the influences of relative humidity and treatment on stratum corneum elasticity, Bioeng. Skin, 2:203-214, 1986.

47. de Rigal, J., and Leveque, J. L., In vivo measurements of the stratum corneum elasticity, Bioeng. Skin, 1:13-23, 1985.

48. Leveque, J. L., Grove, G., de Rigal, J., Corcuff, P., Kligman, A. M., and Saint Leger, D., Biophysical characterization of dry facial skin, J. Soc. Cosmet. Chem., 82:171-177, 1987.

49. Leveque, J. L., and de Rigal, J., Impedance methods for studying skin moisturization, J. Soc. Cosmet. Chem., 34: 419-428, 1983.

50. Clar, E. J., Her, C. P., and Sturelle, C. G., Skin impedance and moisturization, J. Soc. Cosmet. Chem., 26: 337-353, 1975.

51. Tagami, H., ohi, M., Iwatsuki, K., Kanamaru, Y., and Ichijo, B., Evaluation of the skin surface hydration in vivo by electrical measurement, J. Invest. Dermatol., 75:500-507, 1980.

52. Tagami, H., Electrical measurement of the water content of the skin surface, Cosmet. Toilet., 97:39-47, 1982.

53. Tagami, H., Kanamuru, Y., Inoue, K., Suehisa, S., Inoue, F., Iwatsuki, K., Yoshikuni, K., and Yamada, M.,

Water sorption—Desorption test of this skin for in vivo functional assessment of the stratum corneum, J. Invest. Dermatol., 78:425-428, 1982.

54. Wortzman, M., and Grove, G. L., Assessment of long-lasting moisturizers by skin surface electrical hydrometry. Presented at the 7th International Symposium of Bioengineering and the Skin, Milwaukee, Wisconsin, June 16-18, 1988.

55. Kohli, R., Impedance measurements for the noninvasive monitoring of skin hydration—A reassessment, Int. J. Pharm., 26:275-287, 1985.

56. Spencer, T. S., Anderson, P. J., and Seitz, J. C., Use of a phase angle meter to measure product effects on the skin surface, Bioeng. Skin, 2:153, 1986.

57. Serban, G. P., Henry, S. M., Cotty, V. F., and Marcus, A. D., In vivo evaluation of skin lotions by electrical capacitance and conductance, J. Soc. Cosmet. Chem., 32: 421-435, 1981.

58. Serban, G. P., Henry, S. M., Cotty, V. F., Cohen, G. L., and Riveley, J. A., Electronic technique for the in vivo assessment of skin dryness and the effect of chronic treatment with a lotion on the water barrier function of dry skin, J. Soc. Cosmet. Chem., 34:383-389, 1983.

59. Jacques, S. L., A linear measurement of the water content of the stratum corneum of human skin using a microwave probe. In Proceedings of the IEEE, Denver, Colorado, 1979. IEEE Frontiers of Engineering in Health Care (IEEE, New York, 1979), pp. 180-182.

60. Jacques, S. L., Maibach, H. I., and Susskind, C., Water content in stratum corneum measured by focused microwave probe: Normal and psoriatic, Bioeng. Skin-Newsletter 3, 1981, p. 118.

61. Jacques, S. L., Wilson, D. R., and Maibach, H. I., Profile of water content vs. depth in the stratum corneum, 4th International Symposium of Bioengineering and the Skin, Besancon, France, 1983.

62. Rosencwaig, A., Potential clinical applications of photoacoustics, Clin. Chem., 28:1878-1881, 1982.

63. Cahen, D., Bults, G., Garty, H., and Malkin, S., Photoacoustics in life sciences, J. Biochem. Biophys. Methods, 3:293-310, 1980.

64. Campbell, S. D., Yee, S. S., and Afromwitz, M. A., Applications of photoacoustic spectroscopy to problems in dermatological research, IEEE Transactions in Biomedical Engineering BME-28, 26:220-227, 1979.

65. Pines, E., and Cunningham, T., Dermatological photoacoustic spectroscopy. In Bioengineering and the Skin (R. Marks and P. A. Pyane, eds.), MTP Press, Lancaster, England, 1981, pp. 283-290.

66. Simon, I., Emslie, A. G., Apt, C. M., Blank, I. H., and Anderson, R. R., Determination of in vivo water concentration profile in human stratum corneum by a photoacoustic method. In Bioengineering and the Skin (R. Marks and P. A. Payne, eds.), MTP Press, Lancaster, England, 1981, pp. 187-195.

67. Rosencwaig, A., and Pines, E., A photoacoustic study of newborn rat stratum corneum, Biochem. Biophys. Acta, 493:10-23, 1977.

68. Scheuplein, R. J., A survey of some fundamental aspects of the absorption and reflection of light by tissue, J. Soc. Cosmet. Chem., 11:111-112, 1964.

69. Puttnam, N. A., and Baxter, B. H., Spectroscopic studies of skin in situ by attenuated total reflectance, J. Soc. Cosmet. Chem., 18:469-472, 1967.

70. Puttnam, N. A., Attenuated total reflectance studies of the skin, J. Soc. Cosmet. Chem., 23:209-226, 1978.

71. Comaish, S., Infrared studies of human skin in vivo by multiple internal reflection, Br. J. Dermatol., 80:552-528, 1968.

72. Baier, R. E., Noninvasive, rapid characterization of human skin chemistry in situ, J. Soc. Cosmet. Chem., 29:283-306, 1978.

73. Osberghaus, R., Gloxhuber, H., and Braig, S., Hydagen-F, ein neuer Hautfeuchitigkeitsregulator Methoden und Ergebnisse des Wirkungsnachweises, J. Soc. Cosmet. Chem., 29:133-146, 1979.

74. Triebskorn, A., Gloor, M., and Greiner, F., Comparative investigation of the water content of the stratum corneum, using different methods of measurement, Dermatologica, 167:64-69, 1983.

75. Stanfield, J. W., and Kyriakopolour, A., Substantivity of alpha keri bath oil® (Westwood Pharmaceuticals Inc., Buffalo, NY), 4th International Symposium in Bioengineering and the Skin, Bescanson, France, 1983.

76. Potts, R. O., Guzek, D. B., Harris, R. H., and McKie, J. E., A noninvasive, in vivo technique to quantitatively measure the water concentration of the stratum corneum using attenuated total reflection infrared spectroscopy, Arch. Dermatol. Res., 277:489-495, 1985.

77. Potts, R. O., Stratum corneum hydration: Experimental techniques and interpretations of results, J. Soc. Cosmet. Chem., 37:9-33, 1986.

78. Idson, B., In vivo measurement of transdermal water loss, J. Soc. Cosmet. Chem., 29:573-580, 1976.

79. Nilsson, G. E., and Oberg, P. A., Measurement of evaporative water loss: Methods and clinical applications. In Non-Invasive Physiological Measurements, vol. 1 (Rolfe, P., ed.), Academic Press, New York, 1979, pp. 279-311.

80. Grice, K. A., Transepidermal water loss and transepidermal water loss in pathological skin. In The Physiology and Pathophysiology of the Skin, vol. 6 (Jarrett, ed.), Academic Press, London, 1980, pp. 2115-2146, 2147-2155.

81. Miller, D. L., Brown, A. M., and Artz, E. J., Indirect measures of epidermal water loss. In Bioengineering and the Skin (R. Marks and P. A. Payne, eds.), MTP Press, Lancaster, England, 1980, pp. 161-171.

82. Weldon, A. E., and Monteith, J. L., Performance of a skin evaporimeter, Med. Biol. Eng. Comput., 18:201, 1980.

83. Spruit, D., Interference of some substances with water vapor loss from human skin, Am. Perfumer Cosmet., 86:27-32, 1971.

84. Baker, H., Experimental studies on the influence of vehicles on percutaneous absorption, J. Soc. Cosmet. Chem., 20:239-252, 1969.

85. Weil, I., and Princen, H. M., Diffusion therapy analysis of transepidermal water loss through occlusive films, J. Soc. Cosmet. Chem., 28:481-484, 1977.

86. Seitz, J. C., and Spencer, T. S., The use of capacitative evaporimetry to measure the effects of topical ingredients on transepidermal water loss (TEWL), J. Invest. Dermatol., 78:351, 1982.

87. Berube, G. R., and Berick, M., Transepidermal moisture loss. II. The significance of the use thickness of topical substances, J. Soc. Cosmet. Chem., 25:397-406, 1974.

88. Thiele, F. A. J., Hemels, H. G. W. M., and Malten, K. E., Skin temperature and water loss by skin, trans. St. John's Hosp. Dermatol. Soc., 58:218-223, 1972.

89. Powers, D. H., and Fox, C. A., A study of the effect of cosmetic ingredients, creams and lotions on the rate of moisture loss from the skin, Proc. Sci. Sect. Toilet Goods Assoc., 28:21-26, 1957.

90. Rietschel, R. L., A method to evaluate skin moisturizers in vivo, J. Invest. Dermatol., 70:152-155, 1978.

91. Rietschel, R. L., A skin moisturization assay, J. Soc. Cosmet. Chem., 30:360-373, 1979.

92. Buettner, K. J., The moisture of human skin as affected by water transfer, J. Soc. Cosmet. Chem., 16:133-143, 1965.

93. Blank, I. H., Moloney, J., Emslie, A. G., Simon, I., and Apt, C., The diffusion of water across the stratum corneum as a function of its water content, J. Invest. Dermatol., 82:188-194, 1984.

8

Deodorants

S. Mark Henry

Bristol-Myers Products, Hillside, New Jersey

I. INTRODUCTION

Deodorants are designed to prevent or reduce underarm malodor caused by the action of bacteria on perspiration emanating from apocrine sweat glands. Deodorants which contain aluminum and/or zirconium salts may also act as antiperspirants by inhibiting eccrine sweat gland perspiration. Such deodorants, consequently, are classified as pharmaceutical or "cosmetic-drug" products rather than cosmetic products. Whether drugs or cosmetics, however, efficacy claims must be substantiated.

Consumer acceptance of a deodorant as with any product, is based upon several important properties of the product. These include package design, ease of application, fragrance, color, feel, and of course, efficacy. This chapter addresses the origin and nature of axillary odor, its control by deodorants, and efficacy evaluation of deodorants by state-of-the-art clinical trials.

II. AXILLARY ODOR

A. The Apocrine Sweat Gland

The axilla contains eccrine and apocrine sweat glands. Eccrine glands, found in profusion all over the body, produce the voluminous, watery secretion vital in controlling body temperature. Apocrine glands, the ultimate source of axillary malodor, are concentrated in the axillae, but also occur in the mons pubis, the external auditory meatus, the areola, and the circumanal area. In areas other than the axillae, the apocrine glands may be partially or completely nonfunctional.

Hurley and Shelley [1] describe the secretory portion of the apocrine gland as a coiled tube with diverticuli and connections between the coils. A duct connects the coils with the upper end of a hair follicle, or it may exit directly onto the skin surface. Hurley and Shelley [1] provide in-depth descriptions of the anatomy and physiology of the gland, and discuss the stimulatory effect of adrenergic drugs on apocrine gland innervation.

B. Apocrine Secretions

Unlike the clear, watery, eccrine secretion, the fluid secreted by the apocrine gland is rather viscous and turbid. The pH is 5.0-6.5 [1], averaging slightly less acidic than eccrine sweat. The principal lipid component is cholesterol [2], but nitrogenous substances, fatty acids, carbohydrates, salts, and water are also present [1]. Labows et al. [3] have conclusively demonstrated that androsterone sulfate and dehydroepiandrosterone sulfate, previously reported by Julesz [2], are present in "pure" apocrine secretions.

C. Axillary Bacteria and Odor

In 1947, Killian and Panzarella [4] showed that cutaneous bacteria were involved in the production of the distinctive malodor of perspiration. Shelley et al. [5] later demonstrated that nonodorous droplets of apocrine sweat when incubated with bacteria from the axillary vault yield an odor indistinguishable from that found in malodorous axillae. Hurley and Shelley [1] describe a series of experiments implicating the axillary microflora acting on apocrine secretions as the cause of axillary malodor.

Shehadeh and Kligman [6] later showed conclusively that the most abundant bacterial types, gram-positive staphylococci and diphtheroids, are those responsible for axillary malodor.

They stated categorically that gram-negative bacteria are not responsible for axillary odor.

Labows et al. [7], in 1979, reported that the "heating of odorless apocrine secretion to temperatures in excess of 150°C produces a strong axillary odor." The work of these researchers leaves little doubt that two of the key components of axillary malodor are derivatives of androgen sulfates; i.e., 5-alpha-androst-16-en-3-ol and 5-alpha-androst-16-en-3-one. The latter, a ketone, is the more disagreeable and dominant of the two materials.

Differences in the character of the axillary malodor among individuals are partially attributable to differences in bacterial types and numbers in these individuals. Micrococci and diphtheroids are the dominant bacterial groups. Labows et al. [3] have shown that incubation of apocrine secretion with micrococci results in a sweaty odor identified as isovaleric acid [8]. Incubation with diphtheroids produces an acrid malodor.

III. CONTROL OF AXILLARY ODOR

A. Introduction

The oldest and only completely natural method of controlling axillary malodor is liberal and frequent rinsing with water. The relatively poor effectiveness of this procedure is improved if soap is used. However, the fastidiousness of today's society demands all-day protection. Several very different approaches are available to achieve this goal.

B. Inhibition of Odor by Antimicrobials

Since apocrine sweat remains free of malodor in a sterile environment, it is obvious that a reduction in the axillary microflora should also result in reduced odor. One of the earliest applications of this information was to incorporate bacteriostatic chlorinated bisphenols or other antimicrobials into soap bars. Lathering and rinsing the axilla reduced malodor to very low levels and residues of the soap and of the antimicrobial component provided several hours of additional protection.

With the demonstration by Gump in 1953 [9] that hexachlorophene [2,2'-methylene-bis(3,4,6-trichlorophenol)], a highly effective, broad-spectrum antimicrobial, is an effective deodorant in a powder or stick formulation, the deodorant market took on new dimensions. Soon aerosols and lotions became major entities in the deodorant market.

As the record of hexachlorophene toxicity began to develop with its subsequent ban from cosmetics, alternative antimicrobials were evaluated and incorporated into products with varying degrees of success. Some of these are described below.

1. Antibiotics

Aureomycin, tyrothricin, and neomycin have good deodorant effects [10]. Neomycin, however, is the only antibiotic which has appeared in commercial deodorants. Its effectiveness was first demonstrated by Shelley and Cahn [11]. Concern over possible detrimental changes in the axillary flora and the fear of sensitization and development of resistant bacterial strains resulted in the demise of antibiotics as active ingredients of deodorants.

2. Carbanilides

While a number of carbanilides and salicylanilides have been tested and even used in deodorant soaps, the most widely accepted of these is 3,4,4-trichlorocarbanilide (triclocarban). It is used alone or in combination with triclosan.

3. Triclosan

Triclosan (Irgasan DP-200; Ciba-Geigy) is a broad-spectrum antibacterial agent. Activity against gram-positive bacteria is found at very low concentrations. Triclosan has come to be one of the most widely used antimicrobials in deodorants [12]. The concentration of triclosan most often used is approximately 0.25%.

4. Zinc Phenolsulfonate

The effective deodorant concentration of this compound (p-hydroxybenzenesulfonic acid) is about 2.5%. Zinc phenolsulfonate is frequently used in combination with 3% or more of propylene glycol, which has its own antimicrobial properties at 15% or greater.

5. Aluminum and Zirconium Salts

The antibacterial properties of several aluminum salts were reported in 1957 by Blank et al. [13]. Leyden and Kligman [14] demonstrated the broad-spectrum of antimicrobial activity of aqueous aluminum chlorohydrate using in vitro techniques. When applied to the forearm and occluded, 0.1 ml of a 20% solution suppressed growth of all bacteria.

Thus, the deodorant properties of aluminum salts appear to be due to direct antibacterial properties rather than to the ability of antiperspirant salts to reduce underarm wetness. The acidity of antiperspirant formulas may be a major factor in bacterial inhibition. Deodorants/antiperspirants containing aluminum and/or zirconium salts are often marketed with both antiperspirant and deodorant claims.

C. Odor Masking

Colognes and perfumes have been widely used, particularly in Europe, to mask or disguise axillary malodor simply by overpowering it with a more acceptable scent. Carefully selected perfumes may blend with malodor to provide an esthetic and functional masking. One fragrance house claims its special compositions "surround" odor, making it more acceptable. Others have incorporated antimicrobial essential oils to provide both scent and bactericidal properties. Coriander oil is an example of a natural plant extract which has both (moderate) antibacterial and pleasant fragrance characteristics.

D. Odor Neutralization/Suppression

The pH of the combined axillary secretions averages between 6.0 and 6.8 in adult males and females. This slight acidity fosters axillary malodor. Alkalinization of the axilla with sodium bicarbonate or aluminum hydroxide causes instantaneous suppression of malodor. This suppression continues until the alkali dissipates or reacidification occurs through normal physiological processes.

E. Miscellaneous Odor Inhibitors

While the literature [15] refers to several unique substances such as ion exchange resins or absorbents, none of these have achieved commercial success.

IV. EVALUATION OF UNDERARM DEODORANTS

A. Introduction

In vitro bacteriological techniques for predicting the deodorancy potential of a compound or formulation have been used since about mid-century. However, the lack of reliability of these methods for any purpose other than screening new chemicals has been well documented [16]. In 1954, Gee and Seidenberg

[17] published an in vivo sensory method for the evaluation of germicidal soaps. Whitehouse and Carter [18] described a similar method for deodorant soap bars in which trained judges "sniffed" the axillae and rated malodor on a numerical scale. Henry et al. [19] developed a variation of the procedure in which axillary odor was collected on glass tubes which were capped and presented to odor judges. The technique helped eliminate bias and other problems associated with judge/subject interaction. The sensory procedure used most frequently today has been described by Carabello [20] and McOsker [21].

At a meeting of the American Society for Testing Materials (ASTM) in December, 1981, a Task Group (E-19.04.19) was formed to develop consensus guidelines for the clinical study of deodorant efficacy [22]. The resulting recommended guide, now in tentative final draft, is expected to be published in the near future. The method described below is similar to the ASTM recommended guide, but is based largely on the author's experience, previously published techniques, and protocols developed in conjunction with contract testing facilities.

B. Subject Selection

1. Enrollment Criteria

Subjects may be male or female adults in good health and capable of producing at least moderate axillary odor when not using deodorants or antiperspirants. Subjects whose malodor is too low or too high are not acceptable. It is also advisable to select subjects whose left versus right axillary odor levels are not vastly different. Age does not appear to be an important factor. It should be noted that subjects using systemic antibiotics may not be able to meet the odor requirements. Subjects with any discernible axillary irritation are excluded.

C. Odor Judges

1. Selection

Odor judges must be able to discriminate malodor from other odors which may be present in the axillary milieu; e.g., fragrances, smoke, and miscellaneous metabolic end products. Judges must also be able to estimate the intensity of odor on an arbitrary scale. (See Sec. D.2.)

2. Training

A convenient method to aid in the selection of potential judges is to present candidates with five or six dilutions of isovaleric acid in dibutyl phthalate. The following concentrations (D. A. Withycombe, personal communication) may be used: 0, 5, 20, 72, 277, and 600 ppm. A 25-ml aliquot of each is measured into a 1-oz bottle with an 8- to 10-mm orifice. Candidates must be able to arrange the isovaleric acid standards in the correct order of intensity.

An important part of judge training is participation in an actual clinical odor evaluation. Normally, three or four trained judges are used. Judge candidates act as supernumerary judges and qualify if their scores are deemed to be equivalent to those of the trained judges.

D. Method

1. Restrictions

a. Conditioning Period. Five to seven days before the beginning of a study, enrolled subjects enter a conditioning period in which nothing except ordinary, low-fragrance soap and water may be used in the axillae. No deodorants, antiperspirants, antimicrobial soaps, or medicated products of any type may be used. Subjects must also refrain from swimming during the final 2 days of the conditioning period and for the entire test period. Subjects may shave their axillae during the conditioning period up to the last day, but not during the test period.

b. Test Period. In addition to the restrictions noted above, subjects are requested to avoid spicy foods and personal products containing fragrances during the test period. Excessive use of alcoholic beverages is to be avoided. Subjects must not smoke or use highly flavored oral products for at least 1 hr prior to each odor evaluation. Washing of the axillae is permitted only at the test facility. If subjects bathe, shower, or shampoo at home, they must make every effort to avoid getting soap or shampoo in the axillary vault.

2. Odor Evaluations

The magnitude of the intensity scale used for odor evaluations is a matter of choice. The author uses the 0-10 scale recommended by Whitehouse and Carter [18], with 0 indicating no malodor, 5 moderate odor, 10 strong, disagreeable odor. Judges are instructed to disregard fragrances in the scoring. They are also instructed not to smoke, eat, or chew gum just prior to evaluations and are not permitted to use perfumes, colognes, after-shave lotions, or other products which might interfere with odor evaluations. Personnel other than judges assisting in the study are bound by the same restrictions.

Evaluations are made by each judge (no less than three) independently. A subject approaches a judge with arms held close to the sides. One arm is raised and the judge immediately evaluates the level of axillary odor. After entering the score for that axilla, the procedure is repeated for the contralateral side. The left/right order of evaluation is the same for all subjects.

In the indirect (tube) method [19], the bottom end of a 20 × 150-mm borosilicate glass test tube is placed in each axillary vault and held in place by the subject by holding the arms close to the body.

When directed, the subject vigorously rubs the bottom portion of the tube in the axilla and hands the tube to a study assistant, who covers the bottom portion of the tube with a plastic tube closure (KIMKAP; Kimble Div. Owens, Illinois) and places the tube onto a rack. The other axilla is sampled in the same manner. When both axillae have been sampled, the tubes are quickly passed to two judges, who independently evaluate odor clinging to the tubes. If there are more than two judges, a second axillary sampling is obtained using fresh tubes. If sufficient sample is obtained, as indicated by the judges, the subject is released until the next sampling period. If not, additional samples are taken. Judges evaluate panelists at approximately 60- to 90-sec intervals. Each evaluation is entered on a record sheet which does not show subjects' product assignments or any previous odor scores. After each judging period, the investigator collects all record sheets and distributes new ones for the next odor evaluation.

3. Product Application

a. Preparation of the Axillae. The axillae are washed with tepid water and soap (nonantimicrobial, low fragrance) using a gauze pad or wash cloth. The axillae are then thoroughly rinsed with fresh water and dried with a paper towel. Failure to rinse thoroughly may leave a soap residue which can seriously interfere with the study.

b. Lotions (Roll-Ons). Liquid solutions and suspensions are shaken and applied with a disposable 1-ml tuberculin syringe or equivalent device. Approximately 300 mg is applied to each axilla and spread over the vault with a glass rod, the syringe tip, or by the subject using his or her own finger. As an alternative, the applicator container intended for the market may be used to apply the test material.

c. *Sticks (Solids).* An individual stick is used for each subject. Sticks are weighed before and during application until approximately 300 mg has been applied. Final weights are recorded each day.

d. *Aerosols.* Cans are shaken and sprayed for 2 sec at about 6 in. from the axilla. No more than 50% of the contents of any one can should be used.

e. *Other Products.* For deodorant soaps, talcs, pump sprays, etc., the investigator applies the formulation or product in such a manner as to reflect the intended usage for such product. The methods described above may serve as a guide.

4. Control Wash/Evaluation

a. *Control Wash.* On the first day of the test period, normally a Monday following the conditioning period, subjects report to the testing facility for a supervised control wash. Following the rinse, subjects may be given a clean undershirt to wear for the next 24 hr at the discretion of the investigator.

b. *Control Odor Evaluation.* On the second test day, subjects report to the test facility for a control odor evaluation. Subjects must have malodor in both axillae with a grade of no less than 4 (on the 0-10 scale), as determined by at least two trained judges.

E. Schedule

Day 1

Control wash (see Sec. IV.D.4.a).

Day 2

a. *Control Odor Evaluation* (see Sec. IV.D.4.b). The control wash and odor evaluations are usually conducted between 8:00 and 10:00 A.M. The control odor evaluation serves three purposes:

1. Elimination of subjects with unacceptable odor score; final empanelment
2. Establishment of base line scores for untreated axillae

3. Determination if there are sufficient left/right (pretreatment A/pretreatment B) differences to require using baseline scores as a covariate in the analysis of posttreatment data to adjust for side bias

b. Initial Product Application. A subject whose odor is acceptable moves on to an area where the first application of test material(s) is made. The subject is then given a fresh undershirt (optional) and is instructed to (1) refrain from washing the axilla over the next 24 hr, and (2) report back to the test facility 24 hr later.

Day 3

Application/Evaluation

1. Subjects are given the second application of test material(s) and a clean undershirt, and are instructed to report 12 hr later for evaluation.
2. Twelve-hour odor evaluation is made, and subjects receive instructions to appear at the test facility next morning.

Day 4

Final Evaluation. The 24-hr evaluation is made at this time. Subjects are then dismissed from the study.

V. EXPERIMENTAL DESIGN

Choice of design in deodorant efficacy studies depends on the objective of the study. The number of subjects required may differ depending on the particular design. However, a minimum of 20 and maximum of 40 will normally suffice.

The number of product applications may also vary. With active components such as aluminum chlorohydrate, two applications (one per day) appear to be adequate. With actives that may slowly build up residues on the skin (substantive test materials), up to eight applications (one per day) may be required to obtain maximum activity.

For preliminary evaluations of previously untested test materials, it is recommended that Design 1, treatment versus placebo, be used. The placebo can be a vehicle without actives or no treatment. In all cases, half the subjects receive *treatment* in the left axilla and the other half, in the right axilla.

In Design 2, treated versus treated, the test material is compared with a reference standard of known activity or a competitive product. In this design, the degree of odor reduction is not known, since there is only a historical comparison with the initial baseline data.

Design 2 is the basic incomplete block (BIB) in which two products may be compared with placebo, no treatment, or a third product. This design has the advantage of comparing two test materials and simultaneously getting some idea of the extent of odor reduction.

The reader is referred to the ASTM Deodorancy Protocol for an in-depth discussion of study design.

VI. STATISTICAL ANALYSIS

While there are several possible methods of analyzing the data from a deodorant study, the author has had considerable success using a repeated measures design. The odor scores between the two test materials are compared at each time period. Factors used in the analysis are panelists, treatments, test days, and judges. In addition, the number of subjects experiencing superior performance with each treatment are tabulated and treatment differences are tested by the Wilcoxon signed rank test. If analysis indicates product differences from day-to-day and/or judge-to-judge, separate analyses are made for each day and/or judge.

The first day (pretreatment) odor scores may be used as a covariate in the analysis of posttreatment data to adjust for significant differences a subject's odor scores prior to treatment with test materials.

VII. REPORTING DATA

Since deodorant efficacy tests are based on purely subjective evaluations, data are reported as numerical differences between treated and control (or treated-treated) rather than percent reduction of odor. The differences in odor intensities are calculated for each test period. Comparisons between pretreatment and posttreatment intensities should be avoided, since many factors other than the test materials can affect day-to-day odor levels. For these same reasons, it is not valid to compare efficacy data from two different studies.

VIII. CONCLUSIONS

The procedures described in this chapter for evaluating deodorant efficacy have been successfully used by the author for many years. The resulting data have been used to guide product development efforts and to provide claims support.

Methods usually evolve and are improved as an investigator develops experience. Such has been the case for the method described here. Modifications, improvements, and variations will continue as new information and new needs arise.

The reader is encouraged to supplement this basic methodology with additional details and recommendations to be found in the ASTM Consensus Guidelines for the Clinical Study of Deodorant Efficacy [22].

REFERENCES

1. Hurley, H. J., and Shelley, W. B., The human apocrine gland in health and disease, Charles C Thomas, Springfield, Illinois, 1960.
2. Julesz, M., New advances in the field of androgenic steroido-genesis of the human skin, Acta Med. Acad. Sci. Hung., 25:273-285, 1968.
3. Labows, J. N., McGinley, K. J., and Kligman, A., Perspectives on axillary odor, J. Soc. Cosmet. Chem., 34:193-202, 1982.
4. Killian, J. A., and Panzarella, F. P., Comparative studies of perspiration collected from clean and unclean skins of human subjects, Proc. Sci. Sect. Toilet Goods Assoc., 7:3-11, 1947.
5. Shelley, W. B., Hurley, H. J., and Nichols, A. C., Axillary odor: Experimental study of the role of bacteria, apocrine sweat, and deodorants, Arch. Dermatol. Syph., 68:430-446, 1953.
6. Shehadeh, N. H., and Kligman, A., The effect of topical antibacterial agents on the bacterial flora of the axilla, J. Invest. Dermatol., 40:61-71, 1963.
7. Labows, J. N., Preti, G., Hoelzle, E., Leyden, J., and Kligman, A., Steroid analysis of human apocrine secretion, Steroids, 34:249-258, 1979.
8. Labows, J. N., Human odors—What can they tell us? Perfumer Flavorist, 4:12-17, 1979.

9. Gump, W. S., Hexachlorophene in cosmetics, Drug Cos. Indust., 72:622-716, 1953.
10. Ferguson, E. H., A note on axillary odor, J. Invest, Dermatol., 24:567-568, 1955.
11. Shelley, W. B., and Cahn, M. N., Effect of topically applied antibiotic agents on axillary odor, J.A.M.A., 159: 1736-1738, 1955.
12. Fox, C., Antiperspirants and deodorants, Cosmet. Toilet., 100:27-41, 1985.
13. Blank, I. H., Moreland, M., and Dawes, R. K., The antibacterial activity of aluminum salts, Proc. Sci. Sect. Toilet Goods Assoc., 27:24-28, 1957.
14. Leyden, J. J., and Kligman, A. M., Aluminum chloride in the treatment of symptomatic athlete's foot, Arch. Dermatol., 111:1004-1010, 1975.
15. Alexander, P., Antiperspirants and deodorants, Mfg. Chem. Aerosol News, Dec. 1969, pp. 24-40.
16. Baxter, P. M., and Reed, J. V., The evaluation of underarm deodorants, Int. J. Cosmet. Sci., 5:85-95, 1983.
17. Gee, A. A., and Seidenberg, I. L., Evaluation of germicidal soaps, Soap Chem. Spec., 30:42-48, 1954.
18. Whitehouse, H. S., and Carter, R. O., Evaluation of deodorant toilet bars, Proc. Sci. Sect. Toilet Goods Assoc., 48:31-37, 1967.
19. Henry, S. M., Jacobs, G., and Cotty, V. F., An alternative to direct panelist-judge interaction in evaluating underarm deodorants, J. Soc. Cosmet. Chem., 35:283-295, 1984.
20. Carabello, F. B., Guidelines for the clinical study of antiperspirant and deodorant efficacy, Cosmet. Toilet., 95:33-35, 1980.
21. McOsker, D. E., Determination of human axillary and human mouth odor, Cosmet. Toilet., 101:67-77, 1986.
22. Withycombe, D. A., Consensus guidelines for the clinical study of deodorant efficacy, Cosmet. Toilet., 100:55-57, 1985.

9

Antiperspirants

Anthony J. Parisse

Carter-Wallace Inc., Cranbury, New Jersey

I. INTRODUCTION

It is most probable that since before recorded history, man has sought to overcome the malodorous emanations of his body. There are many references in the Old Testament cited by Cooley [1], which indicate the human effort to cover or eliminate bodily odors through personal hygiene and the application of fragrant compounds. Individuals in the ancient civilization of Egypt lavished themselves with perfumes for the bath, the hair, clothing, bed linens, slaves, and horses [2]. Among the artifacts recovered from Tutankhamen's tomb were ornate alabaster perfume vases decorated with carved lotus buds and mandrake fruit [3]. The art of making the body smell better was evident in all the civilizations from Babylon, Assyria, Persia, China, Greece, and Rome. In many instances, the essences used for deodorization were also alleged to have curative and therapeutic properties as well. Throughout the millenia the efforts to overcome body odor continued to be a cover-up with a more pleasing and more powerful aroma.

It was not until more recently that we have taken more direct steps to reduce underarm odors by restricting bacterial growth, or by restricting the perspiration which provides the bacteria the moist environment which helps to promote bacterial growth. In the late 1880s, a cream product called Mum, which contained zinc oxide, claimed to reduce underarm odor. Because zinc oxide is a mild astringent and antiseptic, Mum was one of the first products known to specifically deal with axillary odors with a treatment that was more than a cover-up.

Antiperspirants first appeared on the market somewhere near the very end of the nineteenth century, or in the beginning of the twentieth century. An aluminum chloride solution sold under the trade name of Everdry, and introduced around 1902, has been credited as the first antiperspirant product to be marketed as such. This was followed in 1908 by a brand called Hush. Several years later, in 1914, another product similar in composition, called ODO-RO-NO was marketed nationally [4]. Although these products were believed to reduce perspiration, it was not until 1916 when Stillians published his observation on the effectiveness of an aluminum chloride solution in reducing excessive perspiration after a few applications [5]. These solutions were wet and runny, but in spite of their lack of cosmetic elegance, sold reasonably well. However, the greatest shortcomings of these products were the deteriorating effect they had on fabric and irritation to human skin. The first recorded effort to buffer the effect of aluminum chloride is documented in Lena Tate's 1921 patent, which claims a composition "consisting of aluminum chloride, alum and borox" [6]. The patent claims were for a composition of matter and no specific claims for buffering, or ameliorating irritation, or reducing fabric damage were made. Reference to irritation is made in the second and fourth paragraphs of the invention's description.

In the early 1930s a cream antiperspirant was introduced which contained aluminum sulfate in a cream base. This product had the esthetic qualities of Mum, but was significantly effective in reducing perspiration in the same manner as the drippy, wet aluminum chloride solutions. However, it still retained the quality of deteriorating the underarm area of clothing worn by the user as well as the attendant skin irritancy. Attempts were made to buffer these systems using urea or glycine. These met with moderate success, particularly under warm conditions when ammonia would be released to neutralize any free anion. Other less acidic forms such as aluminum lactates, formates, and alums were used but with markedly less effectiveness.

In the late 1940s, aluminum chlorohydrate was introduced. This was buffered aluminum chloride complex which was 5/6 basic and had a much higher pH than aluminum chloride. This material reduced skin irritation potential and reduced fabric damage to a substantial degree. The salt could be purchased as a solution or in granular and impalpable powder forms. The product development chemist was able to work with a broader range of formulation vehicles [7], which became evident in the years that ensued with the marketing of many new products and package forms. The primary concern of the consumer is odor control, and if he or she perspires heavily, wetness control comes in as a close second. Fortunately, antiperspirants can do both, as was shown by several researchers in the mid 1950s [8-10], which gave the formulator greater latitude in compounding new products.

II. SWEAT: COMPOSITION, SOURCE, INHIBITION

The main purpose of perspiration, or sweat, is to control body temperature through the evaporative cooling effect. Moisture is always leaving the skin either through transpiration, in which the evaporative process is minimal and only water vapor leaves the skin, and is unperceptible or insensible; and through perspiration, in which there is a heavy perceptible or sensible presence of moisture in the form of droplets on the skin surface. In addition to temperature control, the purpose of perspiration is also to eliminate lactic acid, which is generated through muscular activity, and to maintain moisture in the skin, preventing dryness and loss of elasticity.

Human sweat is generated from two sources, the apocrine glands and the eccrine glands (Figs. 9.1 and 9.2), which were named by Schiefferdecker in 1917 [11].

The eccrine gland is tubular, and forms its secretions in the coil. The secretions are emptied through a duct leading to the skin surface. The apocrine gland, also tubular, sends its secretion through a straight channel formed by a hair follicle, whereas the eccrine gland has no need of a hair follicle. Eccrine glands will function from infancy through adulthood, whereas the apocrine glands begin to function during the preteenage and teenage years. The secretions of the apocrine and eccrine glands are quite different. The wall lining of secretory tubes differ between the glands in that a single cell layer is found in the lining of the apocrine gland duct and a double cell layer is found in the eccrine gland duct [11].

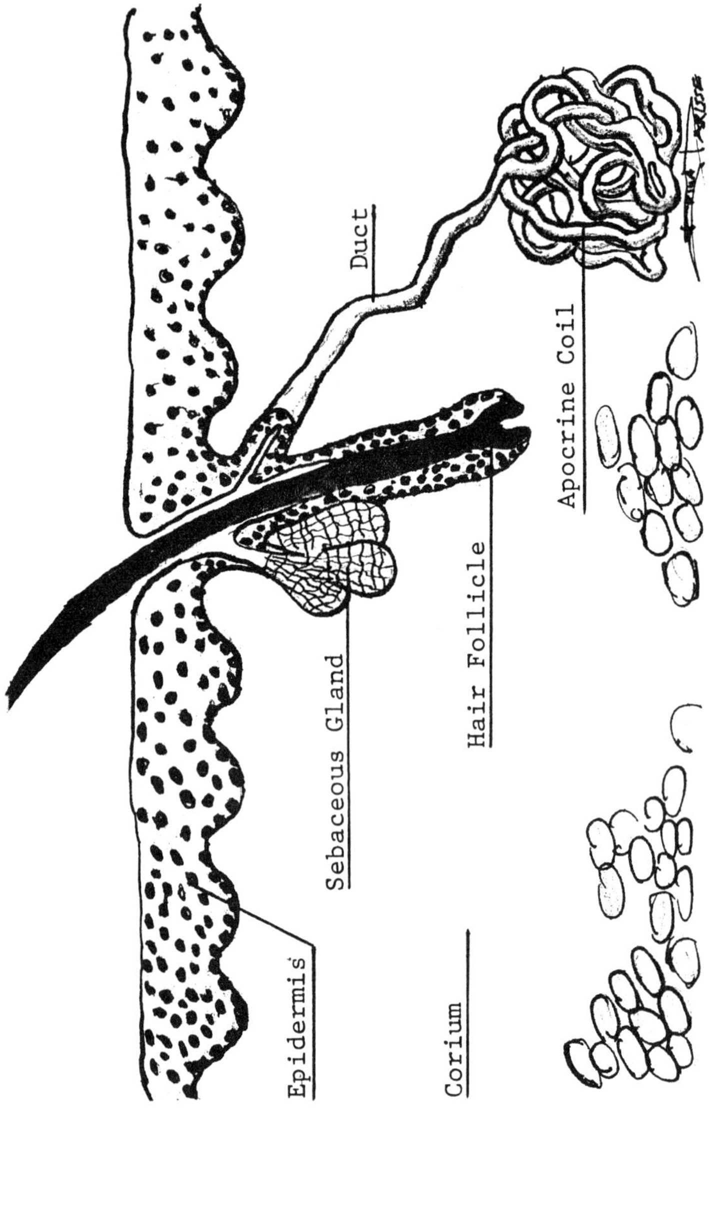

FIGURE 9.1 Diagram of an apocrine gland.

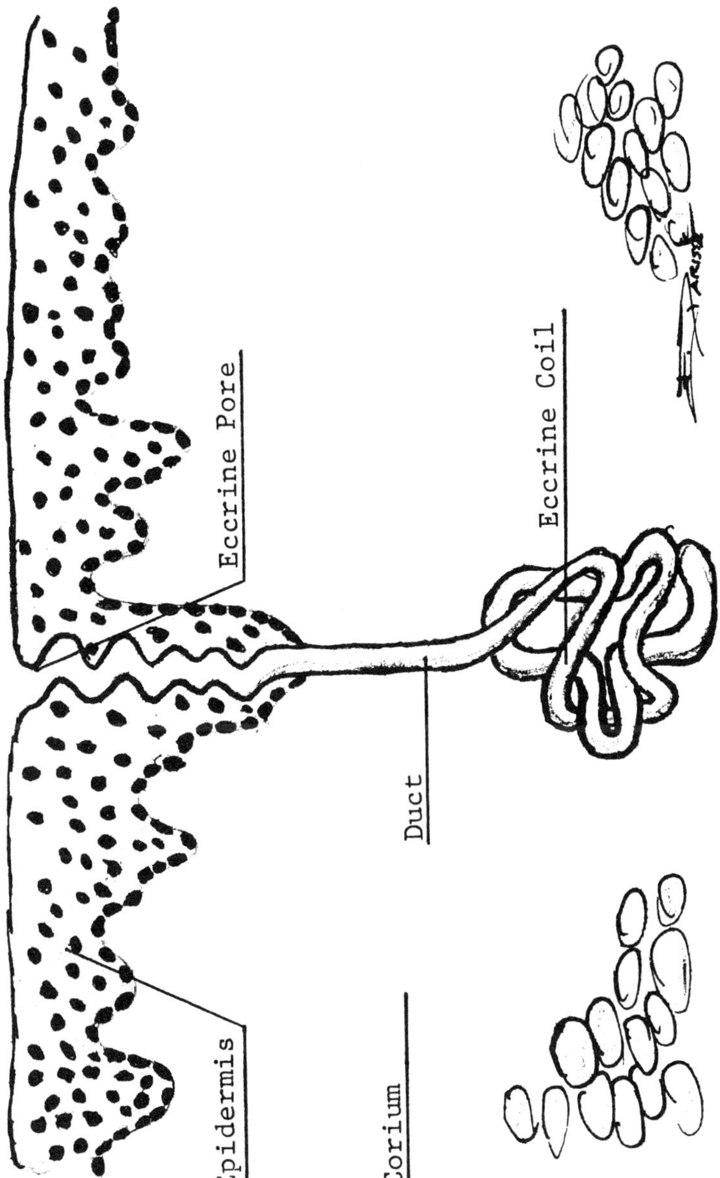

FIGURE 9.2 Diagram of an eccrine gland.

Apocrine glands are found near hair follicles (see Fig. 9.1) with the duct leading from the gland to the hair shaft near the follicle orifice. The apocrine glands secrete a somewhat viscous material containing fatty substances and nitrogenous substances. The secreted liquid can be almost clear to milky white in appearance, and can vary among individuals based on skin color, sex, and diet [12]. Apocrine glands are not equally distributed over the body, but are found concentrated in the axillary, anogenital, and mammary areolar regions of the body [13]. These glands develop after puberty and are not active in children, as it is well known that typical perspiration odor does not occur before maturity.

While apocrine sweat is an important contributor to the malodor generated in the axilla, the perspiration we perceive as a wetness problem is almost entirely that secreted by the eccrine glands. Eccrine glands release large amounts of fluid, chiefly water, in a relatively short period of time, the typical composition of which is given below [14].

Composition of Eccrine Sweat (%)

Water	99.020
Sodium Chloride	0.700
Lactic acid	0.100
Ascorbic acid	0.040
Citric acid	0.040
Acetic acid	0.010
Propionic acid	0.006
Caprylic and caprionic acids	0.005
Urea and uric acids	Trace amounts

Of course the above amounts can vary substantially from person-to-person even within an individual depending on environmental, stress, diet, and many other factors. Other trace components have been found in human sweat; e.g., potassium, calcium, copper, iron, amino acids, and vitamins from the B group [11].

The eccrine glands are found on all parts of the body except the lips and certain parts of the sexual organs. Eccrine sweat tends to be clear with a pH in the higher acidic range.

The eccrine and apocrine glands emit secretions which are relatively odorless; however, we know that odor develops when people perspire. The secretions that carry the fatty exudates from the skin are broken down by bacteria, thus forming new chemical entities in the axillary region, the areas of most concern to us. The bacteria cannot flourish unless a warm moist environment is available. The sweat and the somewhat occlusive nature of the axillar area provide this environment. Thus, the bacterial growth is supported by the moisture and warmth of the underarm, whereas the oily and fatty secretions from the apocrine glands provide nutrients upon which the bacteria can feed. The breakdown of these nutrients is the source of the undesirable odors we perceive [15].

The mechanism by which antiperspirant salts reduce the volume of perspiration provided by the sweat glands has eluded discovery since the first report of Stillians [5] in 1916. The theory that has become generally accepted is based on the work of Quatrale et al. [16], which shows that the perspiration flow is restricted by the amorphous plugs formed within the ducts of perspiration glands below the skin surface. Numerous theories had preceded Quatrale's lengthy and respected research, which include interference of the sweat production at the glandular level [17], resorption of the sweat into the duct before it is released onto the skin surface [18-20], epidermal damage to the skin [21], protein precipitation in the eccrine duct [22], and many other theories, including an electrophysical potential along the sweat duct.

Many investigators still feel that some or even all of these mechanisms contribute to the reduction of sweat in the presence of an antiperspirant, but the strongest evidence currently favors the work previously cited, performed by Richard Quatrale et al. [16]. His work involved a skin stripping procedure using cellophane tape to remove the stratum corneum. In doing so, he was able to induce normal sweat patterns as if no antiperspirant were used. The action of different antiperspirant salts appeared to be different; i.e., whereas aluminum chlorohydrate and aluminum zirconium chlorohydrate were found to be more superficially located in the stratum corneum, the site of activity for aluminum chloride was found to be deeper in the eccrine duct [23].

III. ANTIPERSPIRANT PRODUCTS

A. Ingredients

Antiperspirant ingredients have changed dramatically since the mild, but ineffective, zinc oxide in the 1888 Mum formula, and the very active, but harsh, aluminum chloride used in the 1902 Everdry product. Many metal salts have been explored to find a better, more satisfactory antiperspirant material. Aluminum, chromium, iron, lead, mercury, zinc, and zirconium have been examined [7]. Lead and mercury were immediately ruled out because of toxicological considerations, whereas chromium and iron have coloration and solubility problems. Zinc salts are weakly active compared to the aluminum and zirconium salts. The salts of these two remaining metals are the only ones tentatively approved by the U.S. Food and Drug Administration (FDA) [24]. While this may sound restrictive, the list includes 19 different chemical forms of these metallic salts as follows:

1. Aluminum chlorohydrate
2. Aluminum dichlorohydrate
3. Aluminum sesquichlorohydrate
4. Aluminum chlorohydrex propylene glycol complex
5. Aluminum dichlorohydrex propylene glycol complex
6. Aluminum sesquichlorohydrex propylene glycol complex
7. Aluminum chlorohydrex polyethylene glycol complex
8. Aluminum dichlorohydrex polyethylene glycol complex
9. Aluminum sesquichlorohydrex polyethylene glycol complex
10. Aluminum zirconium trichlorohydrate
11. Aluminum zirconium tetrachlorohydrate
12. Aluminum zirconium pentachlorohydrate
13. Aluminum zirconium octachlorohydrate
14. Aluminum zirconium trichlorohydrex glycine complex
15. Aluminum zirconium tetrachlorohydrex glycine complex
16. Aluminum zirconium pentachlorohydrex glycine complex
17. Aluminum zirconium octachlorohydrex glycine complex
18. Aluminum chloride
19. Aluminum sulfate buffered

The first nine compounds can be used in concentrations up to 25% on an anhydrous basis, and compounds 10-17 can be used in concentrations up to 20% (anhydrous basis). Aluminum chloride is permitted in concentrations up to 15%, calculated in the hexahydrated form. Aluminum sulfate is limited to an 8% concentration provided it is buffered with 8% sodium aluminum lactate.

All the chemical compounds from number 10-19 are restricted to nonaerosol formulations.

The above compounds are not the only salts known to inhibit perspiration, although they are the only ones tentatively recognized as safe and effective by the FDA. Salts of aluminum and zinc that have also been found to be effective include phenolsulfonates, formates, lactates, sulfamates, acetates, and alums [25]. The aluminum compounds listed above are all soluble in water and present little or no problem when formulating in aqueous systems. The chlorohydrates on the other hand do not solubilize as well in alcoholic-based solutions; for that reason, the chlorohydrex complexes were developed to dissolve in anhydrous ethanol, thus facilitating the formulation of aerosol and pump spray products. Generally, many investigators report somewhat poorer antiperspirant effects from the alcohol soluble; i.e., chlorohydrex complex compounds compared to their chlorohydrate counterparts.

Recently a new series of antiperspirant actives has been introduced by the Reheis Chemical Company of Berkeley Heights, New Jersey, and sold under the trade name REACH (Reheis Enhanced Efficacy Aluminum Chloro Hydrate). These compounds are believed to be chemically identical to aluminum chlorohydrate, but have been modified through a new manufacturing process involving a heat treatment under specific conditions. The result being that a physical change occurs in the molecular species of the aluminum chlorohydrate which increases the reduction in perspiration up to 40% more than regular aluminum chlorohydrate. The REACH products are identified as 101, 201, 401, and 501; each with its own degree of effectiveness but all greater than regular aluminum chlorohydrate [26]. These special forms of alunimun chlorohydrate appear to retain their higher antiperspirant activity in nonaqueous formulations such as aerosols. It appears that the REACH species of antiperspirants revert to the lesser effective form when compounded into an aqueous vehicle, or it may be that maximum effectiveness is achieved in water solutions no matter which form of aluminum chlorohydrate is used.

B. Formulations

As indicated earlier, the first formulations of antiperspirants were liquids and creams. The squeeze bottle spray introduced in 1947 under the brand name Stoppette was a packaging development which simply took the liquid form and made it easier to apply. While this form made the product easier to apply, it still had the other attendant problems of a liquid product, the most prevalent being wetness, runniness, and stickiness. Another liquid form was the saturated pad in a glass jar sold under the name of Five Day. While complete coverage could be assured, the product left the user's hands and underarms wet and sticky. Although some additives may have been able to reduce the tackiness of these products, the formulator was limited to the solubility parameters of the vehicles, usually water and alcohol. Increasing the alcohol could increase the solubility of some of the emollient moieties, but would increase stinging potential and reduce solubility of the astringent salts [4].

These liquid systems gave way to the newer more elegant forms that we have today, these being roll-ons, sticks (solids), aerosols, creams, pumps, and several variations of these forms. The roll-on formulations appeared shortly after the squeeze-bottle spray, and rather than use the liquid formula, an emulsion was used to increase viscosity and to prevent dripping and running, and permitted the introduction of emollients to detackify the product as it dried on the skin. A typical emulsion roll-on formulation would be as follows:

Roll-On Antiperspirant (water emulsion)

	%w/w
Water	66.5
Aluminum chlorohydrate	20.0
PEG 40 stearate	5.0
Polysorbate-80	1.0
Cetyl alcohol	3.0
Mineral oil, light	2.0
Glycerin	1.5
Magnesium aluminum silicate	0.8
Perfume	0.2
	100.0

Although the emulsion antiperspirant was more esthetic than the old liquid form, there were still several shortcomings. First, it was still wet and took several minutes to dry; second, it still went through a tacky or sticky stage even though emollients had been added; and third, the evaporation of the water gave a cooling effect which, though not serious, was somewhat uncomfortable. In the mid 1970s, a new anhydrous suspension roll-on formula was developed which was an outgrowth of the concentrate of the aerosol system, which we will discuss below. The nonaqueous roll-on did not feel cool and wet and did not get sticky during the dry-down. A typical formula is shown as follows:

Roll-On Antiperspirant (anhydrous)

	%w/w
Cyclomethicone	74.0
Aluminum zirconium tetrachlorohydrex GLY	20.0
Dimethicone	2.5
SD alcohol 40	1.5
Quaternium 18 hectorite	1.5
Perfume	0.5
	100.0

The anhydrous, or sometimes called the dry-type antiperspirant roll-on, met with a good deal of consumer acceptance, since it not only resolved the problems mentioned above for emulsion antiperspirants, but was also more efficacious because of the more effective astringent salt and the use of volatile silicone, which appears to maximize antiperspirant potential for most astringent salts. However, there was still one drawback, and that was the cost of the raw materials. Both the silicone oil and the aluminum zirconium tetrachlorohydrate glycinate were far more expensive than the water-based systems used earlier. In order to remain competitive with the other product forms, the anhydrous roll-on was sold at a sharply reduced profit margin. To bring the cost of goods down, water was added back into the formula, but only to the extent that it would not affect the advantages gained with the anhydrous system. A formula of this type would look like the following:

Roll-On Antiperspirant (silicone/water base)

	%w/w
Water	46.3
Cyclomethicone	25.0
Aluminum zirconium octachlorohydrex GLY	20.0
Cyclomethicone (and dimethicone copolyol)	7.0
Polysorbate-20	1.5
Perfume	0.2
	100.0

This formula is a compromise from the anhydrous formula using the cyclomethicone as the external phase to give the product the feeling of dryness. It will be interesting to see how well this form does in the marketplace.

The aerosol or atomized antiperspirant has been recommended as the most efficient method of applying an antiperspirant since 1934 [27]. However, squeeze bottles were messy products and formulators tried to prepare various water-based self-propelled systems with little success. The effective astringent salts were only soluble in water and corroded aerosol cans and metallic parts of the valve. Further, the hydroalcoholic systems first developed caused valve clogging.

Antiperspirant salts that were soluble in anhydrous alcohol, e.g., zinc phenolsufonate, were not very effective. Glass aerosol containers could not safely maintain a high enough pressure to atomize the hydroalcoholic sprays adequately. In 1967, a new concept in aerosol antiperspirants was introduced. This product involved using an impalpable powder grade of aluminum chlorohydrate at a low concentration of about 2-5% suspended in an emollient oil and a liquified gas system that had a high enough delivery or spray rate so as to deposit enough astringent on the axillae to be effective [28]. The propellants used at the time were dichlorodifluoromethane and trichloromonofluormethane (P12 and P11). This product was immediately accepted by the consumer, and sales indicated the aerosol form was by far the most preferred. However, in 1973, the use of chlorofluorocarbons was banned based upon the studies of Rowland and Molina, who believed these gases would eventually have a significant effect on the protective ozone layer surrounding the earth in the stratosphere [29]. This forced manufacturers to reformulate their products with liquified hydrocarbon propellants, chiefly propane, isobutane, and butane. The lower density of the hydrocarbon propellants reduced the delivery rate of the valve by about 60%. The concentration of the astringent salt thus had to be increased so that sufficient antiperspirant would be deposited on the skin to be effective. A basic formulation would appear as follows:

Aerosol Antiperspirant

	%w/w
Isobutane	63.0
Propane	12.0
Cyclomethicone	11.0
Aluminum chlorohydrate, impalpable powder	10.0
Isopropyl myristate	3.0
Quaternium 18 bentonite	0.75
Perfume	0.25
	100.00

The hydrocarbons used in this system are highly flammable and present special problems in the manufacture of aerosol antiperspirants. DuPont has developed newer systems utilizing chlorodifluoromethane (propellant 22), chlorodifluoroethane (propellant 142b), and blends thereof [30-32]. Mixtures of these propellants can provide reduced flammability, and are not currently considered a threat to the ozone in the stratosphere. However, these propellants are presently very costly compared to the hydrocarbons and use may be limited depending on the insurance underwriters position on covering liabilities caused by hydrocarbons [33] and state laws concerning the pollutant effect of hydrocarbon in the lower atmosphere. Methylene chloride [34] and dimethyl ether (DME) have also been proposed as alternative propellants, but the former is suspect as a carcinogen, and the latter is just as flammable as the hydrocarbon and presents an odor problem.

Deodorant sticks were popular for many years as gelled soap emulsions containing a large quantity of alcohol and an antimicrobial agent. The earliest antiperspirant stick used alcohol and wax with an astringent salt, but never met with complete commercial success [35]. It was not until the early to mid 1970s that a truly acceptable antiperspirant stick or solid was developed. This product form contained an impalpable grade of antiperspirant powder suspended in a gelled stick containing a wax and using volatile silicones as the liquid phase. These sticks applied smoothly and had almost a creamy texture. An example of a solid antiperspirant is as follows:

Antiperspirant Solid

	%w/w
Cyclomethicone	47.8
Aluminum chlorohydrate, impalpable powder	25.0
Talc	3.0
Stearyl alcohol	23.0
Cetearyl alcohol	0.4
Polysorbate 60	0.3
PEG-150	0.2
Steareth-20	0.1
Perfume	0.2
	100.0

There are other variations of the sticks, some using polymer gum bases, some aqueous emulsion bases. There have been several clear gel systems proposed which are microemulsions or "ringing gels." The formula variants are too numerous to mention here.

When the aerosol formulation was threatened in the early 1970s by the fear of the ozone depletion, another form, the pump spray, emerged and met with some small success with the consumer. Using a somewhat more sophisticated pumping system than was unavailable up to the time, formulations were developed to maximize the dryness of the spray while keeping stickiness and oiliness at a minimum. The formulas were similar to the earlier liquid antiperspirants, but through the use of ingredients that became available during the growth of the aerosol form, the pump spray became more acceptable than the old squeeze bottle sprays. The following is an example of a pump spray antiperspirant:

Pump Spray Antiperspirant

	%w/w
SD alcohol 40	55.5
Aluminum zirconium chlorohydrex GLY	17.0
Cyclomethicone	12.3
Phenyl dimethicone	10.0
Water	5.0
Perfume	0.2
	100.0

Pump sprays still do not approach the dryness of the aerosol spray, nor do they provide the degree of protection of the roll-on and sticks. Further, the package itself has some deficiencies; these being leaking when placed in other than an upright position, valve clogging when left in disuse over several days, and the need to prime the pump with several actuations before obtaining sufficient product with an adequate spray.

Cream formulations, as indicated earlier, have been around a long time. The current market share is small compared to any of the forms mentioned above. Those who still use it seem to like the cosmetic elegance of the cream, the high efficacy it imparts, and the nonirritating, emollient effect it has. While not readily apparent to the consumer, the cream formulations have gradually changed over the years to better more stable emulsions of higher cosmetic quality, typified by the system shown below.

As mentioned before, creams have to be put on with the fingers, and there is an uneasiness about the sanitary state of the container as the consumer repeatedly uses the product. Additionally, there is the problem of vapor loss, resulting in drying out of the product if the cap is not replaced securely on the container. This problem is more serious with this

Cream Antiperspirant

	%w/w
Water	62.0
Aluminum chlorohydrate	25.0
Cetyl alcohol	5.0
Glyceryl stearate (and) PEG-100 stearate	5.0
Dioctyl succinate	2.5
Perfume	0.5
	100.0

product form because the flat round container generally used with cream formulations has a large surface area exposed to the air compared to the volume of product in the jar, and permits a high percentage of water to evaporate even during the short exposure times when the product is used.

Antiperspirant products classified as "lotions" can vary significantly in formulation, ranging from relatively simple clear hydroalcoholic systems to more complex emulsions and emulsion suspension systems. Few, if any, lotion products are sold without some application system such as a roll-on or pump spray. These systems may contain any of the currently acceptable antiperspirant salts in a broad spectrum of vehicles. Because formulations can vary greatly depending upon the preferences of the chemist, it is difficult to write a prototype formula which would typify this category. The reader need only be aware that the formula type has been recognized as a separate and distinct form by the FDA in its tentative monograph, and is considered as a different product type compared to roll-ons, aerosols, solids, pumps, and creams.

There are many variations of the above formulations, which represent only basic systems. Each formulation differs from the other depending on the package in which it will be used, the consumer segment for which it is intended, and the manner in which it is intended to be used. There are liquid sticks, clear sticks, clear gels, clear emulsions, water-in-oil, and oil-in-water emulsions all producing the same effect, i.e., antiperspirancy. The latest information on antiperspirant formulations can be found in the Deodorant and Antiperspirant Documentary [36]. While the above package forms will be around for some time to come, we can expect to see modifications of the package design and new packaging systems, almost all of which will use similar formulations to the ones described above. Some systems will possibly be hybrid formulas, looking to present the advantages of two or more of the current systems while avoiding the disadvantages.

All the aforementioned antiperspirant forms reduce perspiration, but not necessarily to the same extent. The OTC panel appointed by the FDA studied this topic and found that there were indeed differences in antiperspirancy, and reported this in the tentative final monograph issued October 10, 1978 [24] as follows:

Product	Range of effectiveness (%)
Roll-on	14-70
Aerosol	20-33
Solids (sticks)	35-40
Pumps (liquid)	15-54
Creams	35-47
Lotions	38-62

Since the monograph has been written some changes in antiperspirancy of these products have taken place. The range for roll-ons today is really between 40 and 70% reduction. Even the poorest of present roll-on products exceeds 40% reduction. The upper limits of roll-on antiperspirancy is generally achieved with the aluminum/zirconium derivatives in a volatile silicone base. Aerosol antiperspirants have been improving with newer vehicles; namely volatile silicones [37], and implementation of the new REACH astringent salts. It is now possible to easily achieve 45% perspiration reduction with an aerosol formulation. Solids with aluminum/zirconium salts and the new REACH salts can readily exceed 40% reduction and even go into the low 50s. Liquid formulations for pumps are generally in the lower portion of the range given. Pump sprays usually require a large amount of alcohol to yield a spray dry enough for consumer acceptibility. This requires the use of alcohol-soluble antiperspirant ingredients which are most likely to be less active than their water-soluble counterparts. Creams have not changed dramatically over the years, and because they represent a small market segment, not much has been done to improve the efficacy, although it is most likely that the range could be extended to the mid 50-60% range with the use of aluminum/zirconium salts and new vehicles. The range for lotions has not changed significantly, since the given range is broad enough to cover the category.

An understanding of the ingredients and the formulations of antiperspirant products will enable the clinician to better understand the testing procedures and the results obtained therefrom when conducting antiperspirant studies. This knowledge can help explain anomalies and abnormal errors when the studies are analyzed.

IV. METHODS OF ANTIPERSPIRANT TESTING

Although it is not official and has only been proposed as a guideline, the generally recognized method for determining sweat reduction is through human testing; however, there have been many other techniques other than the one given in the FDA guidelines. Several of them will be discussed briefly. Only the quantitative human axillary techniques will be discussed in detail, since they represent the procedures most widely recognized in the industry.

A. In Vitro and Animal Testing

Govett and deNavarre [7] reported a protein precipitation test as a "quick qualitative-semi-quantitative test for astringency." They used a 50/50 solution of preserved egg albumin in distilled water. Taking 5 cc of the solution, they added 1 cc of a 10% test solution at room temperature. From this mixture a precipitate developed which varied considerably depending on the astringent tested. The speed at which the precipitate formed and the character of the precipitate varied greatly, necessitating a good deal of interpretation on the part of the experimenter. Each test material was rated for its precipitative effect by the use of plus signs ranging from one to four to indicate the degree and speed of precipitation. Formaldehyde, which will stem perspiration fairly well, did not show much of an effect in this study. On the other hand, the study did show that aluminum chlorohydrate was equivalent to aluminum chloride in its effectiveness.

Christian and Jenkins [38] reported that egg albumin yielded "uncertain and nondifferential results" in their precipitation studies. After screening a number of proteinaceous substances, they felt that casein was the most uniform in both its quality and in the results obtained. Their procedure involved taking 5 g of casein, suspending it uniformly in 50 cc of 0.1 N solution of HCl, adding additional HCl, and heating the mixture on a steam bath with agitation until the solution becomes clear. The solution was cooled to room temperature and brought up to 100-cc volume with 0.1 N HCl. An aliquot of 5 cc of this solution was combined and shaken with 1 cc of an astringent solution. The concentrations of the astringent solutions varied from 3 to 25%. The mixture was placed in a 37°C incubator for a 15-min period, after which the mixture was cooled and then centrifuged. Some of the supernatant liquid was withdrawn from the centrifuged material, and it was treated with three drops of a 25% solution of aluminum chloride. As in Govett and deNavarre's [7] experiments, a plus sign rating was given to the resulting mixtures except that the scores were in reverse. In the casein precipitation method, a clear solution was given four pluses, whereas a very heavy precipitant in the treated supernatant liquid was given four negative signs. (Govett and deNavarre did not use negative signs.)

There were two discrepancies between the studies. Aluminum chlorohydrate showed very little activity even at the 25% concentration in the casein study, whereas it was as effective as aluminum chloride in the egg albumin study. Aluminum sulfate produced better results in the egg albumin study compared to the casein study. These studies were performed based on the early theory that precipitated protein in the ducts of the eccrine sweat glands stemmed the flow of perspiration by either forming a plug in the duct or decreasing the permeability of the cell walls. However, the procedures did not always produce consistent results, and there was no correlation to the actual perspiration reduction in the human axilla. In order to avoid the variables involved in human testing and better approximate more realistic use conditions, animal models were developed.

Urakami and Christian [39] developed a technique using frog membranes and radioactive tracers which was based on work performed earlier by Lux and Christian [40]. This method involved treating one side of the skin on the back of a frog with a control preparation and the other side was treated in a similar manner with the test preparation. The area covered for each treatment was 6 cm^2 and a 250-mg test sample, if it were a cream product, or 0.3 cc of a lotion or liquid was applied to the test sites. If the product were a liquid or too fluid to remain on the test site, a 2 × 25-cm double-thick gauze pad mounted on waxed paper was attached to the frog's skin. The treatments remained on the skin for 1 hr and then washed off. The skin was excised from the frog and divided into the separately treated portions. The membranes were attached, by means of a thread, to glass tubing 12 mm in diameter and 15 cm long with the ends flared to keep the membranes firmly in place. The membranes were kept moist throughout the procedure and were mounted, inside out; i.e., with the treated side facing the hollow of the glass tube. Two cubic centimeters of water were filled into the tubes, which were placed into 4 cc^3 of radioactive iodide solution (activity of 10,000 counts/min/0.1 cc). After 1 hr and 20 min, a sample of the water from inside the glass tube was withdrawn at four 40-min intervals for activity determination to see how much iodide ion penetrated the frog skin. A mean ratio of the activity of the four standard samples to that of the test samples was determined for each frog membrane. Seven to 10 different frog membranes were averaged for each test. The standard used was a cream product, a description of which was not disclosed in the procedure.

This procedure only could express the results in comparison to the standard cream and did not relate to any degree of effectiveness in humans. It also presented the problems associated with dealing with radioisotopes. While it may have provided a screening technique for product development, it was time consuming and required a good level of competence to perform accurately.

Another animal model was discussed by Lansdown [18] in which he induced sweating in rat foot pads with pilocarpine and inhibited the perspiration with aluminum chloride and aluminum chlorohydrate. His work was predicated on the premise that rat paw sweat glands are similar to human eccrine sweat glands, and that both the human and rat glands can be stimulated by injecting pilocarpine and other chemical agents as well as being suppressed by atropine and other anticholinergics. Male Wistar rats between 180 and 300 g in weight were anesthetized with ether. The food pads were swabbed with an ether-alcohol mixture to defat the tissue and the animal was subsequently injected with pilocarpine at various levels to produce different levels of hydrosis. The rats were restrained so the pads of their feet faced upward. The feet were placed into tubes fitted with rubber shrouds to prevent the evaporation of the sweat and observed with a dissection microscope. The time for sweat to appear on the foot pads was measured on all pads. A starch-iodine solution was used to facilitate observance of the sweat formation on the foot pads. Antiperspirant agents at various concentrations; e.g., aluminum chloride at 3, 6, 12.5, and 25% levels, and aluminum chlorohydrate at 25%, were applied to the right foot pads of three sodium pentabarbitol-anesthetized rats. The left feet were used as untreated controls. The differences in the mean sweating times indicated the relative degree of sweat inhibition in the rat's foot.

Again the results of the rat foot procedure were simply a comparison of various antiperspirant products, giving little correlation to the degree of activity to be expected when applied to humans. While it was easier to control the environment and diet of the rats used in the study, the procedures used for inducing, collecting, and measuring sweat were more complicated than would be for humans. Further, sweating was induced by chemical means rather than thermal or psychological stress, which are the conditions under which humans use antiperspirant products.

B. Human Testing

1. Subjective Methods

Sweat has been measured in humans in ways too numerous to discuss in detail here. Probably the first technique for measuring perspiration was the testimonial method in which a formula was put into the axillary vault under one arm and the other underarm was left untreated. After several hours or days the individual would report whether he had noted any reduction in perspiration. If enough individuals felt that they were drier on the treated side, the formula was considered to be an antiperspirant, as was done in Stillians' report on 20 subjects in 1916 [5].

A slightly better technique involved the use of a number of subjects, who were treated under one axilla with a given amount of antiperspirant for one or several sequential days. When sufficient time had elapsed after the last treatment (at least 1 hr), a judge, who was blind to the treatment assignment, would feel each axilla for wetness by touching them with the finger tips and indicating which side felt wetter or drier. The data were analyzed by comparing the number of times the treated side felt drier compared to the untreated side. More than one judge could be used to obtain a higher degree of precision. A somewhat more sophisticated technique was used in the late 1950s by a few investigators. This was also a touch method, but rather than comparing the underarms for wetness and simply deciding which side was wetter or drier, an estimated score, using a scale that could range anywhere from a three- to a 10-point rating, was given to each underarm. The score given depended upon the degree of wetness perceived by a judge blind to the treatments. Many evaluators preferred to use several judges and average the scores. The scores could be averaged and analyzed statistically. Because the numbers obtained were determined subjectively, the results were at best semiquantitative, and actual percent reductions could not be determined and the test provided only comparative data.

2. Earliest Reported Techniques

In 1934, Kuno [11] described several different methods for measuring perspiration flow. One method involved weighing the subjects on an elaborate Sauter balance and determining by the differences in weight lost the amount of perspiration emitted. Another method suggested required the use of iodine indicators to color the sweat as it appeared on the skin to make it easier to count the droplets formed. He also suggested using a small capillary tube inserted into the sweat pore and recording the amount of sweat observed in the rise of moisture in the tube. This method only measured the output of one pore at a time, and the author indicated it could not be done easily, it could not be done on most pores, and it could not be done on a large portion of the population. Kuno also discussed a perspiration chamber in which he indicated the control of temperature, humidity, and ventilation to measure cutaneous perspiration. Dry air was pumped through tubes to a celluloid dish. The celluloid dish was round when applied to any body surface except for the forehead where it was oval to conform to the dimension of the forehead. The celluloid dish was presumably kept in place by the use of adhesives or adhesive tapes. The dry air picked up the moisture from the skin surface and traveled through a second set of tubes to preweighed U tubes packed with dry calcium chloride to absorb the moisture from the air current passing through them. The U tubes were removed and reweighed.

3. Forearm Methods

Richardson and Meig [41] utilized a technique similar to that of Kuno [11] using forearms as the test sites, but they indicated that the apparatus could be modified to be used on the axilla. They clamped a metal cup to the forearm and a current of air, dried by passing through a Drierite-packed U tube, was passed over the skin and run through another Drierite-packed U tube. The second U tube was preweighed, and thus collected the moisture imparted to the previously dried air by the skin. Two such apparatuses were used simultaneously to compare the untreated site to the treated site. The test subjects were placed in a room heated between 105° and 110°F at a 30-35% relative humidity for a period of 36-45 min before the cups were clamped into place. The air was subsequently passed over the skin for 10-min periods.

Subjects were prescreened with 4 to 10 measurements to assure that both forearms perspired equally. The test material was applied to one forearm daily except for weekends, and measurements were made daily just before each application. The procedure was conducted for a period of 3-4 weeks. The results were expressed as a percent effect which was determined by subtracting the milligrams of sweat collected from the untreated arm from the milligrams of sweat collected from the treated arm, dividing the difference by the milligrams of sweat collected from the untreated arm and multiplying by 100. A negative percent effect indicated a reduction in perspiration, and a positive percent effect indicated a properspirant effect.

This technique while simple and quantitative, only allowed the use of a limited number of subjects, since the operation of the apparatus was limited to one at a time. While several devices could be constructed, a technician could only operate two or three at a given time. The accuracy was reported as being within 10%, which using techniques to be described below could be obtained by simpler more direct methods.

The forearm is perhaps the easiest part of the anatomy with which to work. It is preferred to the back because it has a greater number of sweat glands per unit area, and materials can be wrapped in place and secured more easily on this area. Some investigators still prefer to use this method to at least obtain preliminary or screening data even though the axillae are still the sites for the end use of antiperspirants. Until recently there was little or no correlation of the axilla to forearm sweat output and the effects antiperspirants would have upon these areas. In a paper by Holzle and Kligman [42], the authors tested 16 metallic salts, 10 aluminum salts, and six commercial products. The products were applied to the forearm via occlusive Duhring chambers fitted with Webril cloth, and secured to the limb with wrap-around tape for 3 hr. On the following day, the test subjects were placed in a small room heated to 55°C and 30% RH until they began to perspire. The test subjects were then removed from the room and allowed to sit quietly for a total of 3 hr. Sweat output was then determined by taking a silicone imprint and estimating the number of bubbles formed by the sweat trapped in the hydrophobic film. An estimate of the number of bubbles was made visually, but the authors advised that greater accuracy could be obtained by counting the droplets with an image analyzer. This procedure was compared to an axillary procedure, as described by Fredell and Read [55], and the authors indicated that while higher percent sweat inhibition was seen for the forearm tests, the differences found between products were consistent when compared by the two techniques. The advantages of this procedure were that as many as 16 materials could be tested at once, fewer test subjects were needed, underarm hygiene was of no concern, and the time required to perform the test was shortened. However, to perform a test of this nature, a technician would have to be well trained to take suitable imprints for sweat-suppression estimates. In order to get good estimates of the sweat suppression, an image analyzer should be used, which is a costly piece of equipment if one uses it solely for perspiration tests.

4. Electrical Methods

O'Malley and Christian [43] described procedures using a cup method with dry gas, but the water vapor collected over the skin was measured using an electrolytic cell whereby the electrical conductance of the water dissolved in a methanol-acetone-oxalic acid mixture was monitored by measuring the resistance with a Wheatstone bridge. Bullard [44] also used electrical conductance to determine the relative humidity of air passed over the skin surface in an enclosed area. In his measurements, he used lithium chloride-coated elements in a hygrometer to measure changes in the relative humidity rather than the electrolytic cells for measuring the dissolved moisture in the solutions.

James [45] took the electrical procedure outlined above a step further and used a heating chamber to induce sweat output in the axillae and measured the moisture, absorbed by the dry air passed over the skin, by electrical conductance. The subject was kept in a supine or recumbent position with his arms kept away from the trunk by placing his hands on a trapezelike device. The subject was placed in a incubation chamber where the heat could be controlled. Perspiration flow was initiated by having the subject move his legs in a bicycling motion. The flow was reduced by bringing the subject to a complete rest inside the chamber.

5. Cryogenic Technique

Jenkins et al. [46] used a heated room with axillar cups strapped into place, with the supine subject placing his hands behind his head. Again dry air was passed over the skin, but collections and measurements were made in an aluminum collection coil which was connected by Tygon tubing to the exit port of the axilla cup. The moisture absorbed by the dry air passed through the cup was deposited in the aluminum collection coil by placing the coil in a dry ice/alcohol bath yielding a temperature of -60°C. After a 15-min interval, the collection tube was disconnected from the tubing and capped at both ends. The collection coils, which were previously tared, were dried and reweighed to determine the amount of sweat collected. The test required a good deal of skill, manpower, and expense to perform. Accuracy of the test was not reported in the paper, but it appears that the present procedures will yield similar results with the same degree of accuracy with significantly less effort and equipment.

6. Indicator Techniques

Daley [47] reported that the back could be used just as efficiently for determining antiperspirancy as the axilla. He compared the underarm collection pad method with the back method. Rather than collect the sweat from the back, a cream, treated with finely ground sodium carbonate and phenolphthalein, was applied to the back surface, and as the droplets of perspiration dissolved the sodium carbonate, a pink color evolved from the phenolphthalein indicator. The number of droplets were counted and untreated areas were compared to adjacent treated areas. Using this approach, the author felt that up to six products could be measured at a time, recovery time for using the same sites for another test was much shorter, and the back area lent itself to a better statistical test design.

Kuno had described the use of different dyes to highlight individual perspiration sites [11], and the procedures were elaborated by Maniula, who used bromophenol blue [48] and Prussian blue [49] as indicators to reveal the sweat outlets. The procedure involved greasing Ping-Pong balls with white petrolatum into which the powdered dye was uniformly mixed, and covering the greased balls with gauze. The balls with the dye were left in the axillary vault for 10 min, after which time they were removed and the colored impressions were used for comparison of several treatments.

Herrmann et al. [50] attempted to improve upon this procedure using the same bromophenol blue/sodium carbonate/corn starch/gum tragacanth powder mixture used by Maniula, but coating Whatman No. 2 filter paper rather than a Ping-Pong ball. Standards were developed by using the powdered filter papers pressed on known quantities of water dispersed over sandpaper and the skin of cadavers. The tests were subsequently performed on live subjects, and the amount of sweat was determined by comparing the test powder-treated filter paper to the standard papers and judging which standard quantity the test paper best matched.

As late as 1975, a dye technique was reported by Shelley and Hurley [51] in which a starch-iodine system similar to the one described by Minor [52] was used to illustrate the effects of aerosol sprays on sweat reduction. Dyes or color indicators have many shortcomings; the first being that they only approximate the amount of perspiration observed, and the test involves a significant amount of judgment by the investigator, which will always result in some bias. The dye methods cause staining of the skin and often are difficult to wash off, usually requiring bathing or showering to remove the color satisfactorily. During the study the dyes will stain clothing and often require special washing procedures to remove the stains. The dyes also do not always react uniformly, and for one reason or another fail to show a full sweat pattern.

7. Desiccator Methods

In the methods described above, there was a method in which evaporated sweat in a stream of air was absorbed onto a preweighed desiccant in a U tube. In 1964, Wooding et al. [53] described a method in which perspiration was directly absorbed in a specially designed desiccator taped to the back of human subjects. Thus, the sweat excreted within the enclosed desiccator would be absorbed by the silica gel. The desiccator was a press-fitted aluminum pan with a cover, and was perforated on both sides, and each desiccator had its own outer cannister for weighing in the dry and exposed states. The desiccant used in the study was a blue indicator-treated silica gel, which could be reused by simply heating the uncovered desiccator in an open dish at 350°C until the original blue color returned. The preweighed desiccators were applied to the skin and masked completely with surgical tape to prevent absorption of moisture in the atmosphere or from the unoccluded skin surrounding the test site.

A more detailed procedure using the desiccator was given by Steed [54], in which he strapped the desiccator tins to the body side of the axilla as close to the top of the axilla as possible. The desiccators were weighed, taped into place in the axillae about 9:00 AM, and removed 3 hr later. The subjects treated themselves with the test product(s) at home. The purpose of the exercise was to evaluate the test products under normal conditions of usage. In this study, pretreatment ratios were determined during which time neither axilla was treated; the treatment during which one axilla was treated with the test product; and posttreatment where the treatment was discontinued and the residual effects of the antiperspirant materials were observed.

Although the desiccator technique would appear to give the most reliable data, since the perspiration could be collected under near-normal conditions and the desiccators were in intimate contact with the sweat-producing site, there were a few drawbacks in using the technique. Although the desiccators were light in weight and reasonably small in size, they were somewhat uncomfortable to wear for extended periods of time. Also, since the perforated tins had to be taped into place, the tape had an occlusive or at least a semiocclusive effect on the skin, presenting an artificial environment for the collection of the sweat. Further, the tin presented only sites that allowed sweat absorption from those areas only where the perforations contacted the skin and perspiration under the metal would not be absorbed uniformly. Another problem was that the desiccator covered only a small portion of the axillary vault and did not give an overall measurement of the axillar sweat output. While not overcoming all of the above difficulties, the absorbent pad has had sufficient success to be used almost exclusively today as the test procedure of choice.

8. Absorbent Pad Methods

One of the first references to using absorbent pads was made by Fredell and Read [55] in 1951, who used a packing material of absorbent paper folded to form a 3.25×4.50-in. pad. This pad was backed with a rubber substance usually used for dental dams. The rubber backing and absorbent packing material were held together by wrapping them with one layer of gauze exposed to the absorbing side. The pads were coded according to subject number and right or left axilla and weighed before placement in the axillary vault. The rubber-backed pads were placed into 4-oz glass jars with screw caps and identified to to correspond to the pad held within. The samples were weighed to the nearest milligram.

The pad would be placed in the subject's axilla and held in place by the pressure of the arms against the body with the rubber backing side away from the skin. Several pairs of pads were used with each subject each day, thus obtaining samples at different time intervals. One week was devoted to collecting control data and determining right to left ratios. The second week the subjects applied the antiperspirants to the left axilla and collections were taken. The third week the subjects applied the test material to the right axilla and absorbent pads were again collected. On the fourth week, sweat collections were taken on untreated axillae as was done in the first week.

In their study Fredell and Read [55] concluded that using their methodology antiperspirancy could be determined under normal conditions, and that exercise, humidity, and/or temperature control were unnecessary. The sweat output of the axillae could be measured easily and comparisons could be made directly in evaluating antiperspirant activity. Their data also demonstrated that there were differences in sweat output between the right and left axillae of humans, and that the difference was reasonably consistent within the subject.

Some researchers preferred to use some type of harness to hold the pads in place. Wooding and Finkelstein [56] indicated the use of a harness or tape in the studies they conducted. Dress shields made by Klinert had also been used by several researchers, and were very adequate for retaining the pads in place. The dress shield was of greater value when the study was being conducted under ambient conditions, leaving the subject free to perform routine daily tasks, provided they were not overly strenuous as to cause the dress shield to be displaced. In a controlled hot-room environment, the pad retainer became unnecessary, since the activity of the subjects was greatly reduced and the high temperature and humidity reduced the time the pads had to be kept in place. Under these conditions the subjects merely had to keep their arms at their sides to hold the pads in position.

Majors and Wild [57] described in detail a hot-room procedure in which Webril pads (Kendall Company of Walpole, Massachusetts) were used without dress shields or other retaining devices. The procedure they used in their tests very closely approximated the guidelines later set forth by the FDA in 1982 [58] based on the earlier proposed monograph of 1978 [59]. Their procedure utilized a 1-week minimum abstention period, a 100°F temperature 35% humidity controllled environment, 40 min conditioning period, and two twenty-min sweat collection periods, as later defined in the FDA guidelines. In these studies, comparisons were made for left to right differences, and ambient versus hot-room conditions.

Majors and Wild [57] drew some important conclusions from their studies, most of which are still considered valid by many investigators today.

1. Although the rate of sweating differs among individuals, the percent reduction of sweat by an antiperspirant is an independent function.
2. The axilla on the side of the body with the hand a subject generally uses will usually produce a higher amount of sweat than does the corresponding axilla.
3. The response to the effectiveness of an antiperspirant is essentially a normal distribution.
4. Ambient and hot-room procedures yield similar antiperspirant results.
5. The use of sweating ratios provides more uniform data than the sweating rates on multiple sweat collections from an individual subject.
6. The use of sweat ratios results in a narrower confidence interval in the statistical analysis than does the simple use of sweat rates.
7. There is a great deal of variation in the subject-to-subject response, and it is not consistent with formula variations.

Although many still agree with the above conclusions, there was, and still is, some disagreement among researchers.

Jungermann [60], within the same year, also conducted studies using a sweat ratio technique and compared ambient versus controlled environments. His conclusion differed from that of Majors and Wild [57] in that he felt ambient conditions were far more superior to the controlled environment. Wooding and Finkelstein [56] compared the ratio method to what they called a sides subject effect model, or SSEM. This procedure did not require corrections for the left-right differences. These left-right differences were taken care of in the balance of the design and the randomization procedures used. Their paper concluded that while sweat ratios may reduce the variability of the test, the estimate of the percent reduction is questionable. The SSEM also used log transforms to bring the data into conformance of normality, since it was indicated in an earlier paper [61] that the response to antiperspirant efficacy did not necessarily follow a normal distribution.

Wooding [56] may contradict Majors' [57] opinion on the normality of antiperspirant efficacy, but pragmatically, the distribution of the individual sweat reduction is close enough to normality, the differences obtained in using either statistical methods are indeed small, and the sacrifice in accuracy is not important compared to the overall range within the 95% confidence level. Also, while using the ratio method may present a bias, since the pretest ratio is not constant from day-to-day, there is good evidence to show that the left-right difference does exist, and that adjusting the treatment data with the ratio does increase the precision. When the two procedures were compared, using each analytical technique, it was found the differences in percent sweat reduction ranged from 1 to 6%, which was less than the range of error for the test. Because the arguments were strong for either procedure, and because the difference in the results were within the experimental error, the OTC review panel accepted either procedure as valid; i.e., the ratio method or SSEM. However, the panel chose to select the method of statistical analysis, which will be described later.

V. CURRENTLY ACCEPTED ANTIPERSPIRANT PROCEDURES

The procedure submitted to the U.S. Food and Drug Administration by the OTC review panel was published in the *Federal Register* on October 10, 1978. At that time, it was published as a proposed rule; but almost 4 years later, after considering several comments made on the proposed ruling, a tentative final monograph was published on August 20, 1982, establishing the testing procedure as a guideline. Because it is a guideline, the procedure need not be followed to the letter and changes can be made to accommodate variations in formulations or product claims; however, such variations should be based on good scientific principles and judgment, and one should be prepared to submit the data and results and defend the protocol if and when the monograph becomes a published rule. The following procedure covers all the criteria set forth by the monograph and includes notes, details, and comments by the author to help make the procedure practicable and understandable.

A. When to Test a Formulation

It is, of course, obvious that any new formulation would have to be subjected to an in vivo test before it can be assured that it is indeed effective. Even though companies print a full disclosure of ingredients on the labels of their products, if one chooses to imitate the products of his or her competitors identically, an antiperspirant test is in order to be certain that the product has in fact been duplicated, the method of preparation has not impaired effectiveness, and that the competition itself has prepared an effective formulation. Although it may appear as if a formula has been copied exactly, it is possible for aluminum salts to have greater potency with newer technologies in treating these salts [62]. Chemically these antiperspirant ingredients seem to be the same as the lesser effective material, but because of special techniques used in their preparation or processing, their ability to stem the flow of perspiration is increased to a significant degree. Thus, it is possible to copy a formula exactly as listed on a competitor's label and fail to meet the minimum requirement to be labeled as an antiperspirant, while the competition's product surpasses that requirement.

It becomes more difficult to decide whether to run an antiperspirant test when a formula known to be effective is marginally changed. Some seemingly minor changes can effect the outcome of an antiperspirant test to a significant degree. It has become generally accepted by the industry that changes in color or fragrance of an antiperspirant product has little effect on its efficiency in reducing underarm perspiration. However, minor changes in a solvent system, and particularly the nonvolatile liquid in an aerosol antiperspirant, have been known to cause serious changes in the product's antiperspirancy [63,64]; changing the ratio of one ingredient to another in a given formulation can result in a loss of perspiration protection that a consumer might perceive. Hydrophobicity and lipophobicity can have undesirable results on the formula's efficacy by either chemical or physical means; e.g., lipophobic materials can react to render the aluminum in a salt inactive, and hydrophobic ingredients can act as a physical barrier, preventing the astringent salt from acting on the skin. Whenever there is a reasonable concern over the effect of a formula modification, whether it is an addition, deletion, substitution, or a change in concentration of ingredients, antiperspirant testing is in order.

Sometimes processing changes can affect physical characteristics of an antiperspirant. This can happen when solid antiperspirants are subjected to different processing temperatures, thereby making them harder or softer as the case may be, and affect application rate. Roll-on formulations may be affected by emulsion characteristics which raise or lower viscosity. These variations could raise doubts on the efficiency of the products and require confirmation of their effectiveness.

Finally, packaging changes could present problems in the product's effectiveness. A common example is the aerosol antiperspirant in which a valve change can alter the spray rate, or more subtly, the spray pattern, to reduce the antiperspirancy of the product.

B. Selection of Test Subjects

Test subjects must be free from any dermatological condition in the axillae which might preclude them from participating or completing an antiperspirant test. It is also prudent to include only those subjects who are in reasonably good health, are not pregnant, and are not taking medication that would affect perspiration patterns. The panels should be pretested to determine whether they produce enough perspiration in 20-min periods to participate. Although the FDA Guidelines do not require it, a minimum of 100 mg of perspiration/20 min after an initial 40-min warm-up period is desirable. Some laboratories require 200 mg. The panel must be representative enough so that the difference between the subject with the greatest perspiration output must exceed the subject with the least perspiration output by 600 mg of sweat/axilla in a 20-min period.*

Before participation in his/her first test, the panelist should be given a physical examination by a qualified physician. This physical could be given annually to those subjects who regularly participate in antiperspirant testing. (Some laboratories take blood pressure readings and some are known to give abbreviated physicals before each test.) A file should be kept on each panelist which includes all physicals and any adverse reaction, including anything from irritation to light-headedness that might have occurred during earlier tests. Each physical must be

---

*The FDA guidelines do not specify whether this applies to hot-room conditions, ambient conditions, or both. Because it is often difficult to obtain sufficient perspiration under certain ambient conditions, this section may be modified.

signed and dated by the physician. The panelist must be advised to inform the physician and the testing laboratory of any change in the status of his or her physical health during the year between medical examinations.

Each panelist should be given details as to the nature of the test. A complete description of the conditions under which the panel will be subjected should be disclosed. The possibility of irritation or sensitization to the formula(s) being tested should be mentioned so the subject will be aware of any potential problem during the study and report it to the clinical investigator at the time it occurs. It is a good practice to verbally describe the protocol to the panelist and answer all questions the subject may have before, during, and after the study. The test subjects should be given the option of reading the protocol and the test objective to be certain they are fully aware of the procedures and the possible effects they may experience.

When the subject has been properly apprised of the test procedures an informed consent form similar to the model given below should be read and signed by each test participant. Note that each state and municipality has different laws, and the form given below is just an example of how a consent form can be drawn up. The reader is strongly advised to obtain legal counsel in putting together his own consent form.

CONSENT DECLARATION
for an
ANTIPERSPIRANT EFFICACY TEST

NAME OF TESTING LAB:____________________ TEST NO.______

ATTENDING CLINICIAN:______________________________________

I understand the study in which I am to participate is a research test to determine whether or not sample formulations are effective antiperspirant formulas. There will be (No. of panelists) people participating in this study who will all be required to abstain from the use of all underarm products for a 17-day period prior to the test, Saturday and Sunday included. During this time the testing laboratory will provide each subject with an appropriate deodorant that will not affect the outcome of the test. These deodorants are to be used at home for a 17-day period prior to the test. For the day period during the test week, I will come into the laboratory for treatment with the test antiperspirant. I understand the ingredients and finished formulations have been thoroughly reviewed for safety on human skin.

I certify that, to the best of my knowledge, I am in good health, am not pregnant, and am not aware of any reason why I should not be able to participate in this study. If I become cognizant of an unusual reaction to the test product(s) or procedure(s), I will immediately notify (testing lab) at (testing lab phone number) and follow any instructions given me.

The objectives of this study have been fully explained to me and all questions have been answered to my satisfaction.

All information concerning me will be kept confidential with the exception of data to be submitted to agencies of the U.S. Food and Drug Administration and any other agencies, public or private, in which case my name will be withheld unless absolutely necessary.

I understand I will be paid $__________(total amount) for this test at completion. My participation is completely voluntary and I am free to withdraw at any time. If I withdraw, however, I will not be paid.

SIGNATURE OF PANELIST:____________________DATE:__________

ADDRESS:________________CITY/STATE:____________ZIP:______

PHONE NO.: (area code)______________________________

WITNESSED BY:______________________________ DATE:________

Consent forms should be filed with the data sheets used during the test.

C. Instructions for Subjects

In any biological testing, it is impossible to control all factors that affect the results of a test. This axiom is also true for human antiperspirant testing. To help reduce the variables, the subjects must be given instructions both verbally and listed on a sheet of paper to be taken home for referral. The panelist needs to know the importance of observing the rules pointed out, and that any infraction of the rules will mean disqualification from the study and forfeiture of any remuneration agreed to before the test began. The following are a few of the basic rules; each laboratory will see fit to add their own for better control, safety, and convenience.

1. The panelist must not use any antiperspirant products during the 17-day abstinence period or the 5 days of the test except for those products given for use during the period, which are to be applied either by themselves or by the clinician.

2. The underarms of female subjects may be shaven up to 2 days (48 hr) prior to the test week when the test antiperspirants are applied. No shaving is permitted during the test week.
3. It is important that the subject arrive at the designated times and not miss any of the treatments during the test week.
4. During the test week, the axillae are to be washed under laboratory supervision using the soap solution prepared for the test.
5. The test subjects are required to sit during the sweat collection periods, with preweighed pads in the axillae. The subjects should sit erect with both feet on the floor and their arms resting against their bodies. While conversation among panelists is encouraged, emotional or controversial topics should be avoided.
6. Panelists are to follow instructions of the clinicians exactly to avoid disqualification from the test.
7. If a pad should fall or move out of place, whether it is noticed on the panelist's own person or the panelist notices it on another panelist, it should be reported immediately to the clinician.

The rules cannot be overemphasized, particularly when some of the veteran panels become lax in attitudes. It is wise to review the rules occasionally and reissue a copy of them to "replace lost copies."

D. Equipment

The list of test materials given below are essentially the author's suggestions; in many cases, alternatives are or can become available. The list that follows is one that the author feels best suits the requirements in the studies he has conducted and observed.

1. Deodorants for conditioning period. It is convenient to have a selection of deodorants, specifically, nonantiperspirant products, from which the panelists can choose. Some dislike a particular fragrance, some are allergic, and some prefer a particular product form. By letting them choose from a few, there are usually less complaints and reports of irritation. Be sure each panelist has a 17-day supply.

2. Absorbent nonwoven cotton pads [65]. These pads are used for the sweat collections. They measure 105 × 150 mm and

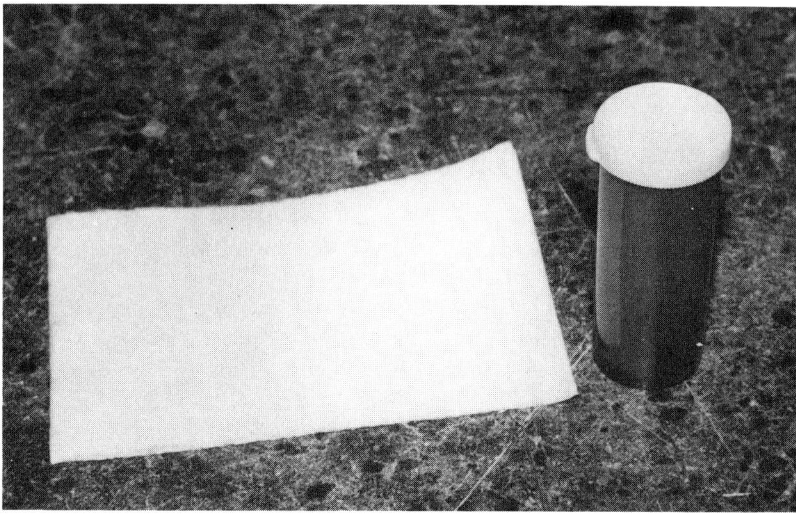

FIGURE 9.3 Unwoven cotton Webril pad and vial container for weighing.

are sold under the tradename Webril. The material can be purchased in large rolls or cut to size (see Fig. 9.3).

3. Pad storage vials. A suitable vial to hold the pad before and after sweat collection is useful. The author has used 1.50" × 3.25" polystyrene vials purchased through local companies or even from the local pharmacy. However, any suitable container will do. Some investigators have used Zip-Lok bags, but care must be taken to be sure they are sealed. The vials have snap caps for easy opening and closing. Whatever container is used, be sure it has a low water vapor transmission rate (see Fig. 9.3).

4. Test Antiperspirant products. The products to be tested should be properly coded with typed or clearly written labels. Reference numbers should be easily traced to the formulation and compounding procedures. Lot numbers, where possible, will give clear and precise reference to the sample's origin. It is a good idea to at least have the active ingredient analyzed in a test lot to be certain that the product being tested has the potency expected. This small, relatively inexpensive step can save a few thousand dollars worth of testing.

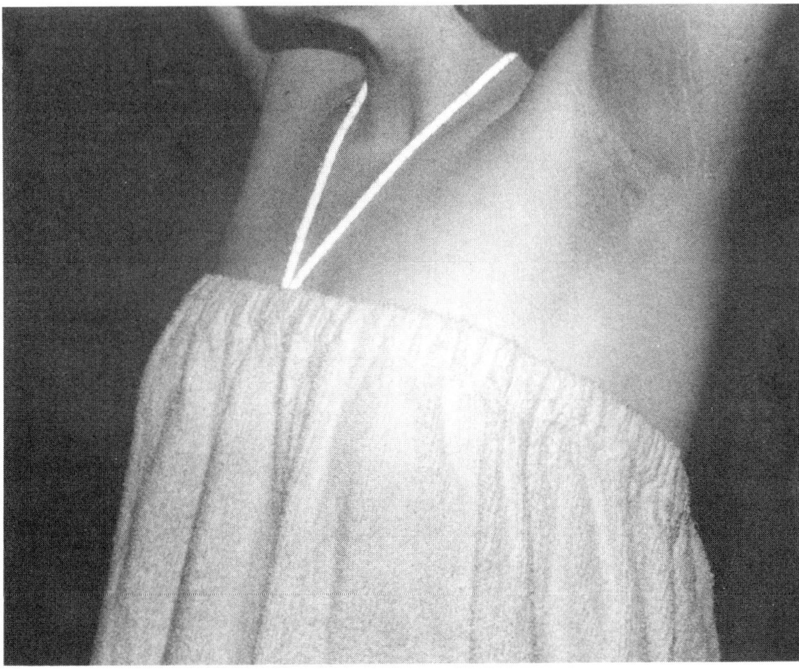

FIGURE 9.4 Hot room cover-up to keep underarms free for pad placement.

5. Apparel for hot room. One of the difficult areas to control in antiperspirant testing is keeping the underarm area free of interference from clothing. Special garments can be made for women which will provide sufficient coverage and leave the underarm area free for the application of the absorbent pads. Figure 9.4 shows such a cover-up, which is made of a cotton terry cloth material to absorb excess perspiration during the study, yet is loose enough to avoid restriction. Of course, this is only one of several designs and fabric materials that may be used for the test. Some women prefer to bring or wear their own garments; these garments should be brought in before the sweat collection day to be sure they will satisfactorily leave the underarm area free for the test.

6. Balance. A dependable, calibrated balance must be used that can determine the mass of the vial and pad to the nearest milligram with an accuracy ±1 mg. Transfer of data

from the balance to a notebook often leads to an error in transposition; it is helpful to have a second clinician checking the weights to reduce this error. Many laboratories are now using electronic balances that have digital displays and can be directly interfaced to a computer to assure the minimal error in transferring of data. Figure 9.5 shows such a system which will print out the weight of the samples before and after the sweat collection and find the difference (amount of sweat collected) with one simple operation.

7. Seats. A seat must be provided for each panelist. The seat's design should be such that the panelist can comfortably sit in an upright position with both feet on the floor. A comfortable padded back rest is needed with slight padding on the seat well. The chair should not have any armrests so the arms can hang loosely against the body.

8. Hot room (for thermally induced sweating). The hot room must be able to accommodate a minimum of 20 subjects based on the tentative monograph guidelines. Temperature is to be

FIGURE 9.5 Automatic electronic balance with cable adaptation for computer interfacing.

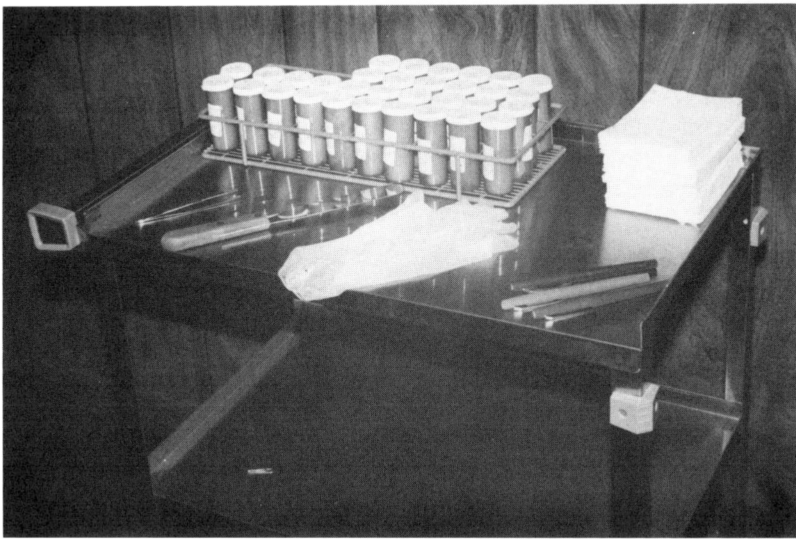

FIGURE 9.6 Laboratory utility cart with racks for vials, spatulas for pad placement, and colored marking pens for vial markings.

maintained at 100°F (±2°F) with a relative humidity between 35 and 40%. Insulation should be sufficient to provide consistent conditions throughout the hot room, and air flow must be adjusted to avoid hot and cold spots. Air exchange is very important with the number of individuals placed in a confined area. Fresh air should be heated and pumped into the room continually. Each individual requires at least 36 ft^3 of space to sit comfortably upright, and additional space is needed for aisles to allow the passage of clinicians and anyone urgently needing to leave the room. The hot room should be monitored with an accurate automatic continuous reading hydrometer and a thermometer with sensing elements in at least three areas of the room to ensure the proper conditions are maintained during the sweat-collection period. Ideally, the conditions should be recorded automatically so that a permanent record of the test condition can be kept. Other amenities may be added, such as carpeting, paneling, and lighting, to make the surroundings pleasant for the panelists who will be spending up to 1 hr and 20 min in the room.

9. Soap solution. A soap solution containing about 1% soap with little or no fragrance dispersed in deionized, filtered water is needed for washing before treatment.

10. Other materials. While not essential, there are several other items which are useful in conducting antiperspirant tests. These include two racks for holding the vials, pharmaceutical spatulas for secure and proper placement of the collection pads, colored felt-tipped pens for differentiating right and left pad vials and numbering same, and a cart for ease of transporting the equipment in and out of the hot room (Fig. 9.6).

E. Hot-Room Procedure

1. Not less than 20 panelists are recruited and instructed to use only the products supplied to them by the laboratory (see instructions for subjects above). After the panelists have satisfied the minimum 17-day rest period, they must come into the laboratory for each treatment day and each sweat-collection day.

2. The test formulation is to be applied to one underarm and the control formulation (placebo) is to be applied to the other underarm on each treatment day.* Application for each treatment must be in the same respective axilla for every treatment.

---

*Note: The FDA guidelines define the control formulation as "...similar as possible to the test formulation and devoid of any antiperspirant activity." The guidelines further state that "Its inactivity is determined in a test compared to no treatment." First, it is often very difficult, if not impossible, to leave out the antiperspirant ingredient in a formula without virtually destroying the formulation. In solid emulsion systems, the active ingredient comprises up to 25% of the formula. Deleting this ingredient will cause the system to fall apart and usually result in an unusable product. In order to restore the formulation to some semblance of its original form, adding other ingredients and revising the ratios of ingredients present in the formula generally leave you with a control formula that is in no way similar to the test formulation. Second, such a procedure requires the formulator to prepare and test *two* formulas, the control and the test formulation. This not only tries the chemist's formulating ability but wastes valuable time, and doubles the cost to test each formula. Third, if the control formula must be compared to no treatment, why should not the test formula be tested against no treatment and avoid the time wasted with two tests?

3. The amount of test product applied must be representative of the average amount applied under normal market conditions by consumers. The product can be applied many ways depending upon the form. Liquids can be applied with a graduated pipette and worked into the axilla with a glass or plastic rod, or rubbed in with a rubber glove or finger cots. Cream can be preweighed on glassine paper smeared into the axilla or scraped off with a spatula and smeared into the axilla (allowance must be made for residue on the glassine paper), and worked into the axilla as above. The product could be applied from the finished package (i.e., aerosols, roll-ons, sticks, etc.) in which case an initial and final weight of the package must be taken with each application to assure proper application amounts.

4. One treatment must be made each day of the test, and all treatments must be recorded and verified. There must be at least one treatment before a test.

5. On the last day of treatment the treated test subjects are asked to enter the hot room. It is best to have the subjects go into the hot room and seat them in the order in which the collection pads will be applied and removed. The test subjects are given unweighed absorbent pads under their arms for a 40-min warm-up period.

6. After the 40 min have elapsed, the unweighed pads are removed and discarded and the preweighed pads are placed under both arms and left there for a 20-min period. The pads are subsequently removed, replaced into their respective weighing containers, and reweighed after the collection period. A second 20-min period is indicated by the FDA guidelines, but only one perspiration collection is required. There is very little difference in perspiration reduction between the collection periods; thus either period should be able to be used for substantiation. There are statistical procedures which permit combining the two periods, allowing use of both sets of data.

7. After the required time is elapsed, the pads are removed and returned to the weighing vials and reweighed within 1 hr after the last pad is placed in its vial. Data must be recorded, verified, and analyzed statistically.

8. Some laboratories prefer to determine the right to left differences or ratios of each subject before the test formulation is evaluated. When this is done, the control formulation must be applied to both axillae of the test subject before incubation in the hot room. If no treatment control is used, no treatment under either arm is used.

F. Ambient or Room Temperature Testing

The FDA guidelines allow for testing under normal or ambient conditions, not requiring an environmental chamber. If this procedure is preferred by the investigator, the above equipment and procedures remain the same except a hot room is not required in the equipment list and steps five, six, and seven of the above procedure would be altered as follows:

5. The treated subjects are fixed with preweighed collection pads inserted to fit well into the axillary vault. The subjects may then proceed with their normal activities (provided unusually strenuous exercise is not included), with particular care taken not to disturb the pads. Some laboratories have used dress shields to prevent the pads from slipping.

6. The panelists are to return for removal of the pads between 3 and 5 hr after application.

7. The pads are returned to their respective weighing containers and reweighed. The data must be recorded, verified, and analyzed statistically.

1. Data Treatment When Pretreatment Data Are Collected

The pretreatment right to left ratios are determined for each subject in the pretreatment collections by simply dividing the amount of sweat from the control axilla into the amount of sweat from the test axilla (the axilla to be treated) (Table 9.1). The posttreatment ratio is determined for each subject by dividing the amount of sweat collected from the control axilla into the amount of sweat collected from the treated axilla (Table 9.2). The adjusted ratio is calculated for each subject by dividing the posttreatment ratio by the pretreatment ratio. The above steps can be demonstrated by the following equations:

$$\text{Pretreatment ratio} = \frac{\text{Test axilla}}{\text{Control axilla}} \tag{1}$$

$$\text{Posttreatment ratio} = \frac{\text{Treated axilla}}{\text{Control axilla}} \tag{2}$$

$$\text{Adjusted ratio} = \frac{\text{Posttreatment ratio}}{\text{Pretreatment ratio}} \tag{3}$$

NOTE: The sweat collected from the control axilla in steps 1 and 2 is collected at different times for each step. Therefore, the amount need not be and most probably will not be the same.

TABLE 9.1 Pretest Data

Subject number	Test axilla (mg)		Control axilla (mg)		Pretreatment ratio
	Right	Left	Right	Left	
1	208			304	0.684
2		116	225		1.009
3		192	237		0.810
4	491			293	1.676
5	1257			1202	1.046
6	836			630	1.327
7	1740			1614	1.078
8	498			301	1.654
9	521			561	0.929
10		222	166		1.337
11	52			38	1.368
12		292	524		0.557
13		295	290		1.017
14		631	528		1.195
15		1306	1396		0.936
16		1115	1210		0.921
17	892			705	1.265
18		78	71		1.099
19	556			264	2.106
20		210	89		2.360
21		926	861		1.075
22	98			121	0.810
23	1421			772	1.841
24		92	145		0.634

TABLE 9.2 Posttreatment Data

Subject number	Test axilla (mg) Right	Test axilla (mg) Left	Control axilla (mg) Right	Control axilla (mg) Left	Pretreatment ratio
1	649			628	1.033
2		216	342		0.632
3		210	334		0.629
4	368			513	0.717
5	819			868	0.944
6	494			650	0.760
7	2408			2063	1.167
8	223			222	1.005
9	196			199	0.985
10		164	249		0.659
11	364			402	0.905
12		491	858		0.572
13		592	613		0.966
14		512	695		0.737
15		877	1394		0.629
16		430	699		0.615
17	436			459	0.950
18		252	463		0.544
19	257			347	0.741
20		224	362		0.619
21		688	590		1.166
22	561			846	0.663
23	189			301	0.628
24		161	331		0.486

The adjusted ratio for each subject is compared to 0.80, which is the ratio that corresponds to a 20% reduction in moisture resultant from a treatment. To test the hypothesis that the reduction in perspiration is greater than 20%, 0.80 is subtracted from the adjusted ratio of all the subjects and the differences are subjected to the Wilcoxon signed rank test [66]. When the median of the adjusted ratio is greater than or equal to 0.80, it is interpreted to mean that less than 50% of the population will obtain a sweat reduction of 20%. When the median of the adjusted ratio is less than 0.80, we can conclude that at least 50% of the target population will experience a sweat reduction of not less than 20%. Using this method, we can expect an error of no more than 5% of the cases where the sweat reduction is not greater than 20%.

a. Example 1. Determining whether an adjusted product can be called an antiperspirant by using pretreatment data (ratio method) and the Wilcoxon Signed Rank Test as recommended by the FDA guidelines. The ratios were determined as follows:

Step 1. The pretreatment ratio is determined by dividing the milligrams of sweat collected from the test axilla by the milligrams of sweat collected from the control axilla.

Step 2. The posttreatment ratio is determined by dividing the milligrams of sweat collected from the treated test axilla by the milligrams of sweat collected from the control axilla.

Step 3. The adjusted ratio is obtained by dividing the posttreatment ratio by the pretreatment ratio.

Step 4. After the adjusted ratios for all the test subjects have been determined, 0.8 is subtracted from the adjusted ratio and the difference (negative or positive) is determined.

Step 5. The rank of these differences is determined by ignoring the signs and ranking the differences beginning with the lowest absolute number and working up. Tie values are averaged in rank as shown in the example below; e.g., since rank 13 and rank 14 have the same absolute numerical value of 0.227, the mean value of 13.5 is used.

Step 6. After the appropriate ranks have been determined, the signs corresponding to the differences are assigned to the ranks.

Step 7. All the positive ranks are added and all the negative ranks are added.

Step 8. The lower value is selected to equal T, as defined in the Wilcoxon Signed Rank Test.

Wilcoxon Signed Rank Test

Subject number	Adjusted ratio	-0.80 = difference	Signed rank
1	1.510	+0.711	+24
2	0.626	-0.174	-10
3	0.777	-0.23	-2
4	0.428	-0.372	-20
5	0.902	+0.102	+5
6	0.573	-0.227	-13.5
7	1.083	+0.283	+16
8	0.608	-0.192	-12
9	1.060	+0.260	+15
10	0.493	-0.307	-19
11	0.662	-0.138	-8
12	1.027	+0.227	+13.5
13	0.950	+0.150	+9
14	0.617	-0.183	-11
15	0.672	-0.128	-6
16	0.668	-0.132	-7
17	0.751	-0.049	-4
18	0.495	-0.305	-18
19	0.352	-0.448	-21
20	0.262	-0.538	-23
21	1.085	+0.285	+17
22	0.819	+0.019	+1
23	0.341	-0.459	-22
24	0.767	-0.033	-3

$\Sigma+ = +100.5$; $\Sigma- = -199.5$; $T = +100.5$.

Step 9. Checking the table for the critical values for the Wilcoxon Signed Rank Test [66], where N (number of subjects) equals 24, we find the probability is greater than 0.10. Therefore, the product is not at least 20% effective in 50% of the population and cannot be considered an effective antiperspirant.

2. Data Treatment When No Pretreatment Data Are Collected

As indicated in the above section, it must be shown that an effective antiperspirant will reduce perspiration by at least 20% in 50% of the population. This is done by multiplying the amount of moisture collected in the control axilla by 0.80. Since half the subjects are treated under the right arm and the other half are treated under the left arm, the adjusted ratios must be determined slightly differently in each instance. The adjusted right/left ratios are then determined as follows:

1. In the case where the right axilla is the treated axilla

$$\text{Adjusted right/left ratio} = X = R/(0.8\ L)$$

2. In the case where the left axilla is the treated axilla

$$\text{Adjusted right/left ratio} = Y = (0.8\ R)/L$$

The data are compared using the Wilcoxon Rank Sum Test [66]. If the median of the adjusted R/L ratio where the right axilla is treated is equal to or greater than the median of the adjusted R/L ratio where the left axilla is treated, the test formula is not considered an effective antiperspirant. If the median of the adjusted R/L ratio where the right axilla is treated is less than the median of the adjusted R/L ratio where the left axilla is treated, the test formula is considered to be at least 20% effective in 50% of the population.

b. *Example 2.* Determining whether a product can be called an antiperspirant by using an unadjusted right to left ratio and the Wilcoxon Rank Sum Test as recommended by the FDA guidelines when no pretreatment data are collected.

Step 1. Organize the data into right side versus left side data and place the treated sides and control sides into subcolumns.

Step 2. Determine the adjusted R/L ratios for X and Y as stated in Eqs. 1 and 2 above.

Collection Data

Subject number	Right side (mg)		Left side (mg)	
	Treated	Control	Treated	Control
1	313			402
2	512			836
3		1087	275	
4	127			301
5	470			421
6		519	362	
7		347	277	
8	496			642
9		426	390	
10		562	227	
11	314			588
12	802			1010
13	153			332
14		448	288	
15		305	201	
16		658	416	
17		544	334	
18		523	431	
19	523			725
20		770	516	
21	613			573
22	305			488
23	1010			1206
24		686	480	

Step 3. Starting with the lowest number first, rank the adjusted R/L ratios disregarding the columns they are in.

Wilcoxon Rank Sum Test

Subject number	R/L (0.8)	Rank	(0.8) R/L	Rank
1	0.973	10		
2	0.766	4		
3			3.162	24
4	0.527	1		
5	1.395	22		
6			1.147	15
7			1.002	12
8	0.966	8		
9			0.874	6
10			1.981	23
11	0.668	3		
12	0.993	11		
13	0.576	2		
14			1.244	18
15			1.214	17
16			1.265	19
17			1.303	20
18			0.971	9
19	0.902	7		
20			1.194	16
21	1.337	21		
22	0.781	5		
23	1.047	13		
24			1.143	14

Step 4. Add the ranks of each column (X and Y) to obtain the sums.

Sum of ranks: X = 107

Y = 193

where

X = the adjusted right-to-left ratio when the right axilla is treated

Y = the adjusted right-to-left ratio when the left axilla is treated

If we let M equal the number of subjects treated on the right side and N equal the number of subjects treated on the left side (both in this case equal 12), we can refer to the tables established for the critical values for the Wilcoxon Rank Sum Test [66] and find that the probability where the sum of the M value is 107 or less is 0.02. Therefore, the product will most probably be at least 20% effective among 50% of the user population, and it can be called an effective antiperspirant.

V. VARIATIONS

The procedure just described can be modified in many ways to satisfy the requirements of the investigator whether it be for convenience, product uniqueness, or claim substantiation. Although the proposed guidelines need not be followed if the procedures are merely designed to acquire information, it is usually wise to adhere to the guidelines should the researcher wish to present the data to outside agencies for review.

A. Three-Day Procedure

Cullum [67] describes a test procedure which takes place over a 3-day period. The subjects are used no more frequently than once every 3 weeks so that there is at least a 14-day abstinence period during which only an alcoholic deodorant product is used. (Note: The FDA Guidelines require a 17-day abstinence period.) On the third week, the subjects are treated once daily for 3 consecutive days with the antiperspirant(s) to be tested. Up to 50 subjects are used per test. After the third application, the subjects are placed in the hot room and the procedures are the same as described in steps 5, 6, 7, and 8 above. The data in Cullum's paper was analyzed two ways using the ratio method and Wooding and Finkelstein's SSEM method. Cullum preferred the SSEM method because it removed the systematic bias.

B. Multiple Test Design

Some investigators may prefer to set up a three-way test design which may include two products and a control or three products, or even one product, an inactive control, and a no treatment control. Such a design should be set up as follows:

	R	L		L	R
Subject 1	A	B	Subject 4	A	B
Subject 2	B	C	Subject 5	B	C
Subject 3	C	A	Subject 6	C	A

where R and L are right and left underarms and A, B, and C are the treatments. Similar designs can also be done for four, five, or more treatments; however, the panel sizes would have to increase to accommodate the additional treatments, requiring larger hot rooms to hold the greater number of subjects

The chief problem with the above design is if one subject does not complete the test the entire block is lost, losing all six subjects, or more, if a greater number of treatments is used. The chief advantage is you can obtain the intercomparison of three treatments or more within one test.

C. Antiperspirancy and Deodorancy Test

Deodorant testing is described elsewhere (Chap. 8); however, it is possible to conduct a single study in which both deodorancy and antiperspirancy data are collected and statistically analyzed. The procedures are the same except they are carried out simultaneously. After the final treatment, the sweat collections are begun 1 hr later and continue for an additional 1 hr and 20 min. After the test subjects leave the hot room, they are given sufficient time to equilibrate with the ambient surroundings (usually about 1 hr and 40 min) and the deodorant test is performed. The data for each phase of the study is analyzed separately.

Performing the study in this way is somewhat more complicated, but with practice it can be performed efficiently and accurately. More important than the cost savings possible with this test is the efficient use of antiperspirant panels which require the long conditioning period prior to the test. Each time a panel is used to test an antiperspirant product for a purpose other than antiperspirant efficacy, the panel is lost for at least 17 days before it can participate in another study whether it be for antiperspirancy or deodorancy.

D. Twenty-four Hour Antiperspirancy

In order to demonstrate all-day protection for an antiperspirant, it is desirable to perform a hot-room sweat collection 24 hr after the last treatment. This is performed in the same manner as described above, but the sweat collection is taken 24 hr later rather than on the same day. Of course a regular sweat collection could also be taken 1 hr after treatment as well as the 24-hr collection.

Another variation would be a 24-hr deodorancy and antiperspirancy test. This becomes logistically impossible, but the sweat collection can be made 22 hr after the last application and the deodorant evaluation made 24 hr after the last application. Again, 1-hr sweat collections and 4-hr deodorant evaluations could be combined with this procedure with the attendant savings in cost and panelists.

E. 130°C Hot Room

It is strongly recommended that clinicians avoid conducting studies involving temperatures above those suggested in the FDA proposed monograph and guidelines for the comfort and safety of the panelists; however, some investigators have used temperatures up to 130°F to simulate the thermal stress that might be encountered in unusually warm environments such as the desert, enclosed vehicles, sitting in a tropic sun, etc. While the amount of perspiration obtained per unit time is greater, similar antiperspirant results can be expected.

F. Emotional Sweating

Researchers have long recognized the emotional state of humans strongly influences both the onset and the volume of sweat. Reller [68] indicated that mental and emotional stimulation play a significant role in axillary sweating. He graphically demonstrates a subject's response to solving arithmetic problems over a period of 120 min by measuring the milligrams of sweat collected at 10-min intervals. The rate of sweat generated by normal stress over a 20-min period remained at about 400 mg/hr, and during mental stimulation reached a peak of about 2400 mg/hr, and sharply decreased back to 800 mg/hr when the stimuli were removed. He also points out that changes in axillary sweat output can change with a change in body positions by measuring sweat output in the sitting, standing, and reclining positions. His paper goes on to show the effects of exercise, leaning against the back or side of a chair, leg position, and salt intake.

Quatrale et al. [69] reported sweat studies conducted on emotionally stressed human subjects. The methods described were mental arithmetic, electric shock, and a word association list. The sweat output was measured using two techniques; the first was electronic hydrometry instrumentation, and the second was the use of the absorbent axillary pads described previously. The authors concluded that the word association list provided the most desirable results, since it was found to be successful in about 75% of the subjects tested. They also preferred the use of absorbent pads because a large number of subjects could be studied simultaneously. Although no antiperspirant formulations were tested, the authors conclude that this method, i.e., the use of the word association with absorbent pad collections, would provide a reliable test procedure to test antiperspirant products for effectiveness under the conditions of emotional stress.

While not appearing in formal journals, some investigators have tried other methods; e.g., having college students keep absorbent pads in the axillae during an examination, and exposing the test subjects to a stressful environment such as enclosed areas containing passive but unattractive animals. The latter procedure tends to provide erratic data and unexpected reactions by the test subjects. The former procedure provides a true situation where the test subjects are under genuine mental stress.

G. Regulatory Status of Antiperspirants

The FDA announced in the *Federal Register* on January 5, 1972, that independent review panels would be assembled to examine the safety, efficacy, and labeling of all over-the-counter (OTC) drugs [70]. A proposed monograph was issued on October 10, 1978 [71], and a tentative final monograph was issued on August 20, 1982 [72]. Although these tentative regulations and guidelines are not presently binding, discretion and good judgment of the major brands in the antiperspirant category have prevailed in adhering to these proposed rules. While some latitude has existed in the formulation and testing of antiperspirants, there has been no attempt to introduce a new antiperspirant ingredient. The FDA would more than likely require a New Drug Application, and demand the product be removed from the marketplace.

The monograph covers the area of safety, testing, labeling, etc., of antiperspirants, and reading it is considered a must for anyone wishing to be informed on the subject. We will look into those aspects affecting antiperspirant efficacy testing only.

One area covered by the OTC review panel was that of user perception tests based on studies performed by Majors and Carabello [73]. The panel recommended and the FDA agreed [24] that in order for an antiperspirant to be considered effective, a majority of users may reasonably expect to see the difference between using an effective antisperspirant and an ineffective one. Whether the FDA considers the Majors and Carabello study sufficient to satisfy this requirement for all antiperspirant formulations, or if they consider that new formulations and/or forms although they have made it clear for other claims to be discussed below. The Majors and Carabello study involved a large population, and at this writing has not been confirmed at another test site, and whether testing of this nature has been continued has not been discussed among investigators.

The question of user perception testing comes up again concerning emotional sweating. Even though data have been presented to the panel demonstrating that emotionally induced sweat can be reduced, there was no user perception data to show that the user could sense the antiperspirant effect. The FDA indicated that until such data becomes available, the claim for emotional sweating will not be allowed. The FDA also adviseed that claims for problem or especially troublesome perspiration will require user perception tests in order to be allowed. In this section, however, they state, "Once the level of activity that is perceivable by users has been established for each of these claims, using the Panel's recommended guidelines, it will no longer be necessary to perform user perception testing on individual products." This clearly points out that a single test would be sufficient to satisfy the testing requirements for these claims. However, the tests would require extremely large panels, and no one to date has taken the steps to satisfy this requirement. A case in point is the problem perspiration claim. The FDA states that a product should be able to reduce perspiration by a greater amount than a standard antiperspirant, and that the test subjects should be heavy perspirers. Heavy perspirers are defined by the panel as the upper 5% of perspirers. In order to assure statistical significance based on the Majors and Carabello report [73], the user perception test would have to be performed on at least 200 subjects for each level of effectiveness to assure a difference exists between the 20% control formulation and the formulation for problem perspiration. If the test formulations are at the 30, 40, 50, and 60% reduction levels, we would have to impanel 200 of the top 5% of the perspirers. In order to recruit such a select group, 4000 subjects would have to be recruited and subjected to sweat testing to obtain the panel. After the sweat testing was completed to screen out the top 5% of heavy perspirers, at least four additional studies at the various sweat reduction levels would have to be conducted. This reason alone is sufficient to deter any single company from collecting such data, since the largest hot room accommodates a maximum of 50 people, and few testing labs have a pool of 4000 subjects to draw from. Even if someone were to perform the necessary work, once the monograph has been issued by the FDA, the claim becomes available to anyone who wishes to use it, since the FDA states "...it will no longer be necessary to perform user perception testing on individual products."

The FDA strongly feels that a control product should be used. They believe that bias can effect the test because of psychological effects. The difficulties in using a control product were discussed earlier in this chapter, the FDA recognizing some of these difficulties states that "The control formulation is similar as possible to the test formulation and devoid of any antiperspirant activity. Its inactivity is determined in a test compared to no treatment." Using the logic applied above, how does one remove the psychological bias in the no treatment when testing the ineffective control formulation? Osborne et al. [64] addressed the vehicle effect on antiperspirant activity and made the following conclusions: (1) The vehicle does not affect antiperspirant activity. (2) The vehicle affects the rate of release of the active ingredients. (3) While the lotion, hydroalcoholic solution, and cream vehicles provide greater than 30% reduction, the stick provides a negative reduction or properspirant effect. (4) The vehicle because of cosmetic appeal and subjective acceptance may influence the antiperspirant efficacy. (5) The side to be treated may be randomly selected without affecting the outcome of the test. Osborne's paper, however, did not present any real evidence to show that subjective responses concerning cosmetic appeal influenced efficacy, and it is also in disagreement with the monograph which stated that antiperspirant sticks provide from 35 to 40% reduction in perspiration. It appears the question of vehicle versus no treatment controls is still unanswered.

Extra strength claims are disallowed by the FDA because the terms imply an increased amount of antiperspirant per dose of product and an additional reduction in perspiration. Since formula variations can affect efficacy, the amount of antiperspirant does not always have a serious effect on the formulation's activity.

VI. NATIONAL ADVERTISING DIVISION AND NATIONAL ADVERTISING REVIEW BOARD

The National Advertising Division (NAD) is a self-regulating system established by the advertising industry around 1971 to monitor claims for products and investigate complaints made by consumers or competitors to maintain truth and accuracy in national television, radio, and print advertising. The National Advertising Division is a division of the Council of Better Business Bureaus, and while it does not have legal authority, it can ask companies to submit data to substantiate claims. If the NAD decides that the substantiation is not satisfactory, it can request the claims be withdrawn or modified.

If agreement cannot be attained between the company in question and the NAD, the controversy can be appealed to the National Advertising Review Board (NARB). The NARB can call upon 50 individuals who represent national advertisers, advertising agencies, and the public. Decisions made by the NARB are final unless additional data or information not reviewed previously become available. A very great percentage of challenges are settled through NAD before they require appeal or referral to the NARB. Many cases are substantiated by the challenged company, and those that are not are usually satisfied by the company altering or discontinuing the offending claim. Data acquired and prepared in accordance with FDA guidelines would be appropriate for review by the NAD.

According to Smithies [74] during the period of 1971-1985, 26 cases concerning antiperspirants/deodorants were brought before the NAD. Only two of these claims reached the NARB for review. Antiperspirants were the subject of 23 cases, which included the stick, roll-on, pump spray, and aerosol spray forms. Claims have involved comparative antiperspirant claims, "apples to oranges" claims (i.e., where two different product forms are compared), implied claims, etc. Thus, the number of cases brought before the NAD over the past 14 years have been small in number relative to the size and scope of the antiperspirant industry. Perhaps it has been the impact of the OTC review board and the tentative final monograph issued by the FDA that has helped keep the industry in check, or it may be that the industry keeps a close check on itself. In any case, standards have been set which most marketers feel they can maintain.

REFERENCES

1. Cooley, A. J., The Toilet in Ancient and Modern Times, Burt Franklin, New York, reprinted 1970.
2. Miller, J. et al., The H&R Book of Perfume, Johnson, London, 1984.
3. Edwards, I. E. S., Tutankhamen, His Tomb and Its Treasures, Metropolitan Museum of Art and Alfred A. Knopf, New York, 1976.
4. Jass, H. E., The history of antiperspirant product development, Cosmet. Toilet., 95:25-31, July 1980.
5. Stillians, A. W., The control of localized hyperhydroses, J.A.M.A., 67:2015, 1916.
6. Tate, L. H., U.S. Patent 1,371,822, March 15, 1921.
7. Govett, T., and deNavarre, M. G., Aluminum chlorohydrate, new antiperspirant ingredient, Am. Perfum. Essen. Oil Rev., 49:365-368, 1947.
8. Aguilar, T. N., Blaug, S. M., and Zopf, L. C., A study of the antibacterial activity of some complex salts. J. Am. Pharm. Assoc., 2nd ed., 45(7):498, 1956.
9. Klarman, E. G., Chemical and bacteriological aspects of antiperspirants and deodorants, J.S.C.C., 7(2):94, 1956.
10. Blank, I. H., Moreland, M., and Dawes, R. K., The antibacterial activity of aluminum salts, Proc. Sci. Sec. TGA, 27:24, 1957.
11. Kuno, Y., Human Perspiration, Charles C. Thomas, Springfield Illinois, 1956, pp. 43; 223-250.
12. Hurley, H. J., and Shelly, W. B., The Human Apocrine Sweat Gland in Health and Disease, Charles C. Thomas, Springfield, Illinois, 1960.
13. Robinson, S., and Robinson, A. H., Chemical composition of sweat, Physiol. Rev., 34:202, 1954.
14. Peck, S. N., Rosenfield, H., Leifer, W., and Biermann, N, Role of sweat as a fungicide, Arch. Dermatol. Syph., 39:126, 1939.
15. Shelly, W. B., Hurley, H. J., and Nichol, A. C., Axillary odor, Arch. Dermatol. Syph., 68:430, 1953.

16. Quatrale, R. P., Waldman, A. H., Rogers, J. G., and Felger, C. B., The mechanism of antiperspirant action by aluminum salts, I, II, III. The effect of tape stripping on aluminum salt inhibited eccrine sweat glands, J.S.C.C., 32:67-73, 107-136, 195-222, 1981.

17. Goodall, McC., Innervation and inhibition of eccrine and apocrine sweating in man, J. Clin. Pharmacol., 20:235-246, 1970.

18. Lansdown, A. B. G., The rat foot pad as a model for examining antiperspirants, J.S.C.C., 24:677-684, 1973.

19. Papa, C. M., The action of antiperspirants, J.S.C.C., 17:789-800, 1966.

20. Papa, C. M., and Kligman, A. M., Mechanisms of eccrine anidrosis. II. The antiperspirant effect of aluminum salts, J. Invest. Dermatol., 49:139-145, 1967.

21. Lansdown, A. B. G., Production of epidermal damage in mammalian skins by some simple aluminum compounds, Br. J. Dermatol., 89:67, 1973.

22. Papa, C. M., and Kligman, A. M., Mechanisms of eccrine anidrosis. I. High level blockade, J. Invest. Dermatol., 47:1, 1966.

23. Quatrale, R. P., Thomas, L., and Birnbaum, J. E., The site of antiperspirant action by aluminum salts in the eccrine sweat glands of the axilla, J.S.C.C., 36:435-440, 1985.

24. Antiperspirant drug products for over-the-counter human use; tentative final monograph, Fed. Reg.. II:36504, August 20, 1982.

25. Plechner, S., Cosmetics and Technology, Wiley Interscience, 2nd ed., 1972, pp. 373-416.

26. Reheis Claims Breakthrough in Aerosol Antiperspirant Efficacy, Soaps, Cosmetics, Chemical Specialties, December 1985, pp. 44-45.

27. Auch, R., Manufacture of Solid Deodorants, Am. Perfum. Essen. Oil Rev., 29:187, 1934.

28. Spitzer, J. G., and Osipow, L., U.S. Patent 3,968,203, 1976.

29. Rowland, F. S., and Molina, M. J., Chlorofluoromethanes in the Environment, Atomic Energy Commission Report No. 1974-1, September 5, 1974.

30. Daly, J., New propellants for aerosol fragrances, Aerosol Age, 29:26-32, September 1984.

31. Crooker, R. M., Alternative propellants; a nonflammable blend especially suited for aerosols in glass, Aerosol Age, 29:31-34, March 1984.

32. Daly, J., Emerging new propellant blends, Aerosol Age, 31:30, October 1986.

33. Rizzo, J. J., The development of the duPont aerosol flammability test (AFT), Aerosol Age, 31:20-25, July 1986.

34. Editor, Methylene Chloride: A White Paper, Aerosol Age, 30:10-12, June 1985.

35. Moore, W. C., U.S. Patents 2,087,161 and 2,087,162, 1937.

36. Deodorants and antiperspirants documentary, Cosmet. Toilet., 100:68-75, December 1985.

37. DeSapio, A. J., New Approaches to Antiperspirant and Deodorant Formulations, HAPPI, February 1986, p. 43.

38. Christian, J. E., and Jenkins, G. L., A comparison of a new astringent agent with such agents now commonly used, J. Am. Pharmaceut. Assoc., 2nd ed., 39:663, 1950.

39. Urakami, C., and Christian, J. E., An evaluation of the effectiveness of antiperspirant preparations using frog membrane and radioactive tracer techniques, J. Am. Pharmaceut. Assoc., 42:179-180, 1953.

40. Lux, R. E., and Christian, J. E., Permeability of frog skin by means of radioactive tracers, Am. J. Physiol., 162:193, 1950.

41. Richardson, E. L., and Meigs, B. V., A method for comparative evaluation of antiperspirants, J.S.C.C., 2:308, 1951.

42. Holzle, E., and Kligman, A. M., Simplified procedure for evaluating antiperspirants: A method for rapid screening with subsequent assessment of axillary antiperspirant activity, J.S.C.C., 34:255-262, 1983.

43. O'Malley, W. J., and Christian, J. E., The design of a continuous recording in vitro method of measuring sensible perspiration over a limited area, J. Am. Pharmaceut. Assoc., 49:398, 1960.

44. Bullard, R. W., Continuous recording of sweating by resistance hygrometry, J. Appl. Physiol., 17:735-737, January-November, 1962).

45. James, R. J., A new and realistic electronic approach to the evaluation of antiperspirant activity, J.S.C.C., 17:749-767, 1966.

46. Jenkins, J. W., Ouelette, P. A., Healy, D. J., and della Lana, C., A Technique for Perspiration Measurement, Toilet Goods Assoc. No. 42, December 1964.

47. Daley, E. W., Antiperspirant Testing: A Comparison of Two Methods, Proceedings of the Scientific Section of the Toilet Goods Assoc. No. 30, December 1958.

48. Manuila, P. L., Traitment medicamenteux local de la transpiration axillaire, Dermatologica, 100:305-309, 1950.

49. Manuila, P. L., and Isler, H., Noveau test de transpiration, Dermatologica, 102:302-306, 1951.

50. Herrmann, F., Prose, P. E., and Sulzbuger, M. B., Studies on sweating. IV. A new quantitative method of assaying sweat delivery to circumscribed areas of the skin surface, J. Invest. Dermatol., 17:251, 1951.

51. Shelley, W. B., and Hurley, H. J., Jr., Studies on topical antiperspirant control of axillary hyperhydrosis, Acta Dermatovener, 55:241-260, 1975.

52. Minor, V., Eines Neues Verfahren zur der Klinischen Untersuchung der Schweissbansderung, Stsch. Z. Nervenh. Eilk, 101:302, 1927.

53. Wooding, W. M., Jass, H. E., and Ugelow, I., Statistical evaluation of quantitative antiperspirant data, J.S.C.C., 15:579, 1964.

54. Steed, M. W., Evaluation of antiperspirant preparations under normal conditions of use, J.S.C.C., 26:17-28, 1975.

55. Fredell, W. G., and Read, R. R., Antiperspirant-Axillary Method of Determining Effectiveness, Toilet Goods Assoc., No. 15, May 1951, pp. 23-27.

56. Wooding, W. M., and Finkelstein, P., A critical comparison of two procedures for antiperspirant evaluation, J.S.C.C., 26:255-275, 1975.

57. Majors, P. A., and Wild, J. E., The evaluation of antiperspirant efficiency—Influence of certain variables, J.S.C.C., 25:139-152, 1974.

58. Antiperspirant drug products for over-the-counter human use, Fed. Reg., 43(196):46,694-746, October 10, 1978.

59. Antiperspirant drug products for over-the-counter human use, Fed. Reg., 47(162):36,492-36,505, August 20, 1982.

60. Jungermann, E., Antiperspirants: New trends in formulation and testing, J.S.C.C., 25:621-638, 1974.

61. Wooding, W. M., The Interpretation of Gravimetric Axillar Antiperspirant Data, Proceedings of the Joint Conference Cosmetic Science, 1968, pp. 92-105.

62. Magres, J. J., and Tykowski, J., Increased efficacy for antiperspirants, Drug Cosmet. Indust., February 1986, p. 51.

63. Geary, D. C., and Krevald, H., U.S. Patent No. 4,151,272, Wax-Like Antiperspirant Stick Compositions, April 24, 1979.

64. Osborne, G. E., Lausier, J. M., and Lawing, W. D., Statistical evaluation of vehicle effect on antiperspirant activity with a limited number of subjects, J.S.C.C., 35:179-191, 1982.

65. MacMillian, F. S. Kilmer, Reller, H. H., and Snyder, F. H., The antiperspirant action of topically applied anticholinergics, J. Invest. Dermatol., 43:363, 1964.

66. Wilcoxon, F., and Wilcox, R. A., Some Rapid Approximate Statistical Procedures, Lederle Laboratories, New York, 1964, pp. 7, 9, 24, 28.

67. Cullum, D. C., A rapid hot room procedure for testing the performance of antiperspirants, J.S.C.C., 29:399-412, 1978.

68. Reller, H. H., Factors affecting axillary sweating, J.S.C.C., 15:99-110, 1964.

69. Quatrale, R. P., Stoner, K. L., and Felger, C. B., A method for the study of emotional sweating, J.S.C.C., 28:91-102, 1977.

70. Fed. Reg., 37FR85, January 5, 1972.

71. Fed. Reg., Vol. 43, 21CFR, Part 350, October 10, 1978.

72. Fed. Reg., Vol. 47, 21CFR, Part 350, August 20, 1982.

73. Majors, P. A., and Carabello, F. B., Presentation to the Advisory Panel on OTC Antiperspirant Drug Products Concerning Hill Top Research Method of Antiperspirant Evaluations, OTC Volume 140065.

74. Smithies, R. H., The NAD/NARB approach to evaluation of antiperspirant and deodorant advertising claims, Cosmet. Toilet., 100:59-63, December 1985).

10

Sunscreen Efficacy

Ward L. Billhimer

Hill Top Research, Inc., Cincinnati, Ohio

I. INTRODUCTION

Since earliest times, people have been exposed to the sun and subsequently needed protection from its harmful effects, but only recently have chemical sunscreens been clinically evaluated.

A. Historical Perspective

Nearly all civilizations have at one time or another included sun worship in their cultures. The ancient Greeks, Romans, Egyptians, Mayans, and Aztecs all revered the sun as a deity who endowed mere mortals with life and health. While using sunlight as therapy for various diseases is quite evident in those civilizations, there is essentially no reference to sun protection as part of their customs.

In the 1700s and 1800s, Western civilization shunned the sun. A peaches-and-cream complexion was the ideal, and almost no one practiced sun bathing. White skin was a sign of affluence and fashion, whereas tanned skin was associated with peasants and the working class. Sun protection during this period consisted of parasols, hats, and nonrevealing clothing.

Sun worship regained popularity in the 1920s when the French designer Coco Chanel made the "bronze" look chic.

Soon, tans became inextricably associated with health, leisure, youth, and international glamor.

As sun bathing became increasingly popular, the need for products with active sunscreen ingredients developed. Salicylates and para-aminobenzoate (PABA) were examined as early as the mid 1920s for their ability to absorb ultraviolet light, and these substances have been used as chemical sunscreens ever since [1]. In fact, the first commercial sunscreen, which appeared in 1928, was a combination of benzyl salicylate and benzyl cinnamate [2]. An old "Tibetan recipe" containing 5% tannic acid was also reported to be quite effective at that time even in the glacial regions [3].

In the 1930s and early 1940s, ultraviolet light-absorbing compounds abounded. Most of these compounds fell into the following categories [4]:

p-Aminobenzoates
Anthranilates
Benzimidazoles
Carbazoles
Cinnamates
Naphthol sulfonates
Pyrrones
Salicylates

Soon after World War II, an obsession with tanning developed, which was further kindled by the introduction of the bikini bathing suit in 1946 by the French designer Louis Reard. As people extended their sun exposure to get faster, deeper, darker tans, sunburn became a more serious concern. Consequently, commercial sunscreens proliferated; many claimed to promote even tanning as well as protection against sunburn.

In reality, most of the chemical sunscreens available in the 1950s and 1960s provided only marginal protection, especially under the stress of prolonged exercise, sweating, and swimming [3-6]. Effective active ingredients such as PABA and its esters were available, but until the late 1960s and early 1970s, the importance of the vehicle to the sunscreen's efficacy was not well understood [5-8].

As numerous investigators have found, the vehicle not only influences the total concentration of active sunscreen on the skin surface, but also affects its skin penetration, water resistance, and the rate and extent to which the treated skin is damaged and then returns to normal [5-10]. Today, the vehicle is considered the second most important ingredient in a sunscreen product [11].

B. Current Regulations

For many years in the United States products containing sunscreens were considered cosmetics because they were labeled solely as tanning products. However, in August 1978, the U.S. Food and Drug Administration (FDA) classified products intended for preventing sunburn and similar conditions as drugs, regardless of claims made for the products [12]. (To date, sunscreens are still classified as cosmetics in Europe, Japan, and Australia.) The FDA based its decision on the definition of "drug," as stated in section 201(g) of the Federal Food, Drug, and Cosmetic Act of 1938:

> ...articles intended to cure, mitigate, treat, or prevent disease and articles intended to affect the structure or any function of the body [13].

Indeed, products containing sunscreen agents affect the structure and function of the skin by absorbing, screening, reflecting, or scattering those portions of the ultraviolet spectrum that cause actinic damage. In addition to claims of preventing sunburn, the FDA also permits claims of protecting the skin from cancer and premature aging [12].

Under the proposed regulations, the FDA compiled a list of sunscreen compounds that are safe, effective, and approved for use in products marketed in the United States. This list appears in Table 10.1 along with the maximum approved concentrations for use. The FDA also proposed methods for evaluating sunscreen efficacy as well as a system for categorizing products in the marketplace based on their performance in preventing sunburn.

The FDA's proposed regulations had a tremendous impact on the cosmetic industry. By allowing claims of protection against skin cancer and premature skin aging, the FDA gave a new dimension to sunscreen use, one of true preventive medicine. Likewise, the FDA had created new opportunities and consequently new life for the industry. The race was on to formulate more effective products (including everyday cosmetics such as moisturizers, creams, lipsticks, and shampoos), to capitalize on the long-term protective capacity of these products, and to identify new broad-spectrum ultraviolet B (UVB) and A (UVA) absorbers.

TABLE 10.1 Sunscreen Agents Approved by FDA for Use in Sunscreen Products

Sunscreen agent	Approved dosage (%)
I. Chemically active	
A. UVA absorbers	
Dioxybenzone	3.0
Menthyl anthranilate	3.5-5.0
Oxybenzone	2.0-6.0
Sulisobenzone	5.0-10.0
B. UVB absorbers	
Aminobenzoic acid	5.0-15.0
Amyl dimethyl PABA	1.0-5.0
Diethanolamine p-methoxy cinnamate	8.0-10.0
Digalloyl trioleate	2.0-5.0
2-Ethoxyethyl p-methoxy cinnamate	1.0-3.0
Ethyl 4-bis(hydroxypropyl)-aminobenzoate	1.0-5.0
2-Ethylhexyl-2-cyano-3,3-diphenylacrylate	7.0-10.0
Ethylhexyl p-methoxy cinnamate	2.0-7.5
2-Ethylhexyl salicylate	3.0-5.0
Glyceryl aminobenzoate	2.0-3.0
Homomenthyl salicylate	4.0-15.0
Lawsone with dihydroxyacetone	0.25 with 3.0 DHA
Octyl dimethyl PABA	1.4-8.0
2-Phenylbenzimidazole-5-sulfonic acid	1.0-4.0
Triethanolamine salicylate	5.0-12.0
II. Physical absorbers	
Red petrolatum	30.0-100.0
Titanium dioxide	2.0-25.0

To date, the FDA has not finalized the over-the-counter (OTC) monograph of sunscreen drug products. Nevertheless, the FDA contends that it may still take regulatory action against products or manufacturers which do not conform to the proposed rule.

II. ULTRAVIOLET RADIATION AND ITS EFFECTS ON SKIN

Although sunlight in general is blamed for sunburn, only a fraction of the sun's rays actually causes the problem. Here on Earth, the solar spectrum consists of wavelengths from 290 to 3000 nm. (The sun does emit wavelengths shorter than 290 nm, but this radiation, known as short-wave ultraviolet, or UVC, is absorbed by the ozone layer of the Earth's atmosphere and does not reach the surface.)

The solar spectrum is generally divided into three categories: infrared (760-3000 nm), visible (400-760 nm), and ultraviolet (290-400 nm). As shown in Figure 10.1, ultraviolet light makes up only about 6% of the spectrum at the Earth's surface (sea level), yet this portion causes nearly 99% of the side effects of sunlight [14].

Ultraviolet light at the Earth's surface is also divided into categories: long-wave ultraviolet, or UVA (320-400 nm), and middle-wave ultraviolet, or UVB (290-320 nm). This UVB radiation—a mere 0.5% of the total solar spectrum—is the portion of sunlight that causes the redness, tenderness, swelling, and occasional peeling or even blistering of the skin known as sunburn.

The symptoms of this cutaneous inflammation or erythema reaction are transient. The skin usually responds to overexposure to the sun within minutes or hours. The erythema reaction generally reaches its peak in 12-24 hr, and then slowly subsides over the next few days. Individual sensitivity to ultraviolet radiation as well as the intensity and length of exposure determine the severity of the reaction.

Unfortunately, even when the redness and discomfort of a sunburn pass, the damage to the skin remains. Years of repeated exposure to the sun can affect the collagen and elastic fibers in the skin as well as reduce the amount of the skin's subcutaneous fatty tissue [15-18]. In essence, the skin ages prematurely [19]. This effect is cumulative and probably irreversible.

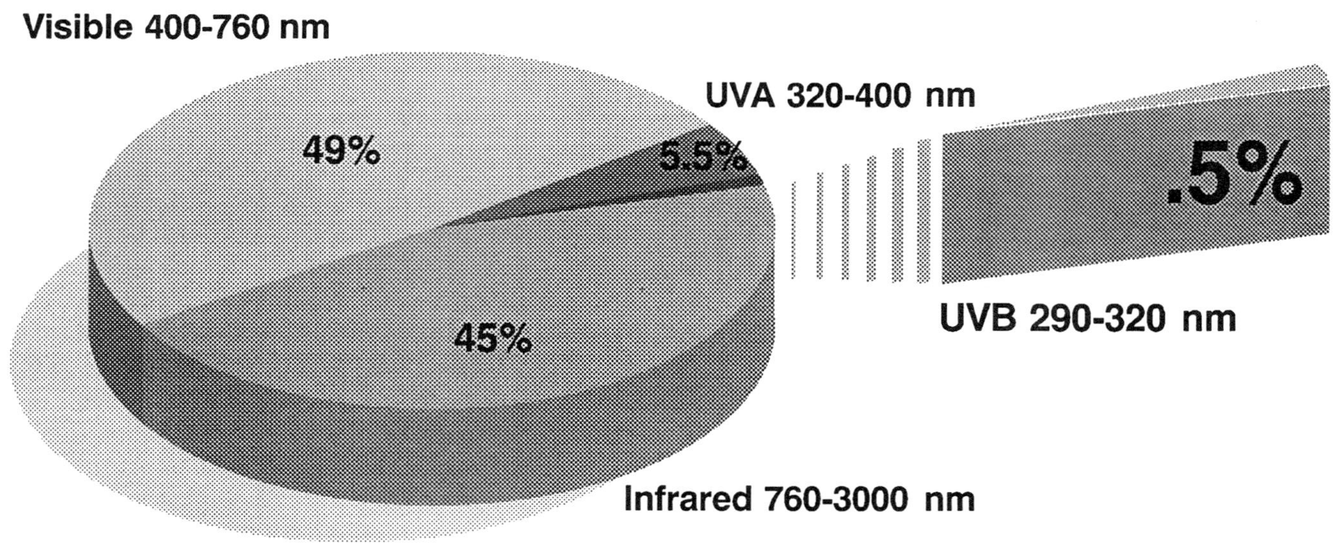

FIGURE 10.1 Approximate composition of solar spectrum at Earth's surface (sea level).

An even more disconcerting result of excessive exposure to solar radiation is the increased probability of skin cancer. Between 1978 and 1986, public awareness of the adverse effects of overexposure to the sun increased dramatically. Through its National Education Program, the Skin Cancer Foundation cited that over 500,000 new cases of skin cancer are reported each year in the United States, and that these cases can be partly attributed to excessive exposure to the sun's damaging rays [20]. The number of reported cases of this disease continues to grow each year. And there seems to be a direct connection between a history of painful, blistering sunburns during childhood and adolescence and the likelihood of developing malignant melanoma [21].

These harmful effects must be minimized. Of course, one way is to limit exposure to the sun. But for many people who work outdoors, or for those who cannot resist the allure of getting tan, limiting exposure is not a viable alternative. In these cases, the best protection from the sun's hazards is using topically applied chemical sunscreens.

A recent study indicates that one-third of the people who use sunscreens select products based on the information on the label [22]. This information must therefore be as accurate and standardized as possible to enable consumers to make informed decisions.

III. THE SUN PROTECTION FACTOR

Sunscreens are intended to protect the skin from sunburn, but since the amount of protection needed varies among individuals, sunscreen manufacturers offer consumers different levels of protection. These different levels of protection are currently expressed as a numerical value, called the sun protection factor, or SPF.

As shown below, the SPF is the ratio between the minimal erythema dose (MED) of sunscreen-protected skin and the MED of unprotected skin.

$$\text{SPF} = \frac{\text{MED of protected skin}}{\text{MED of unprotected skin}}$$

The MED is the smallest amount of ultraviolet light or the shortest exposure time required to produce a minimally perceptible sunburn 16-24 hr after irradiation. For example, if unprotected skin shows a mild sunburn after only 20 min of sun

TABLE 10.2 European SPF Categories Prior to 1978

Sun protection factor	Level of protection	Skin type
3	Minimal	Nonsensitive skin and skin already accustomed to the sun
4	Moderate	Normally sensitive skin
6	Extra	Sensitive skin

exposure, and the same skin protected with a sunscreen burns minimally after only 60 min of exposure, then the SPF of that product would be 3.

The SPF equation evolved from the quotient system introduced by F. Ellinger in 1934 to characterize the efficacy of sunscreen products in preventing delayed erythema [23]. Ellinger's quotient consisted of the erythema threshold time of untreated skin divided by the erythema threshold time of skin treated with a sunscreen. With Ellinger's formula, as the efficacy of the product increased, its numerical value decreased. Ellinger's method was rather crude and inadequately defined; consequently, it was not adopted as a standard method for evaluating sunscreen efficacy.

In 1956, Rudolf Schulze described the degree of ultraviolet light protection afforded by a sunscreen as simply the light protection factor [24,25]. This factor was basically the reciprocal of the ratio described by Ellinger. Studies conducted by Greiter in 1974 and Ippen et al. in 1977 refined this method and coined the term sun protection factor [23,26,27].

The SPF concept was widely accepted in Europe. Table 10.2 lists the principal SPF ratings used there at that time [12].

In its proposed sunscreen monograph of 1978, the FDA recommended adopting the SPF concept being used in Europe; however, the FDA specified somewhat different categories for the SPF ratings, as shown in Table 10.3.

There is often a tendency to regard the SPF of a product as an absolute value. However, this rating is based on a biological assay. In spite of attempts to standardize various aspects of the test used to calculate the SPF, this value is subject to considerable inherent variability. Therefore, these values should be used only for comparison.

TABLE 10.3 FDA-Proposed SPF Categories

Sun protection factor	Level of protection	Skin type
2 to <4	Minimal	Insensitive to normal, rarely burns to burns minimally, tans well to profusely
4 to <6	Moderate	Normal, burns moderately, tans gradually
6 to <8	Extra	Sensitive, always burns easily, tans minimally
8 to <15	Maximal	Sensitive, always burns easily, never tans
15 or greater	Ultra	Sensitive, always burns easily, never tans

IV. CLINICAL EVALUATION OF SUNSCREEN EFFICACY

Despite the development of animal in vivo and in vitro testing procedures, human clinical testing remains the accepted method for determining the SPF values of sunscreen products. In its 1978 monograph on sunscreens, the FDA proposed specific procedures for testing the efficacy of these products as well as guidelines for interpreting the test results. To date, these procedures have not changed in the United States.

A. Executing the Test

The sunscreen test itself is basically a 3-day procedure. A flow chart of the entire process is shown in Fig. 10.2.

1. Selecting Test Subjects

The first step in preparing for the clinical test is selecting a panel of at least 20 volunteer test subjects. These panelists should be healthy adults with no history of abnormal responses

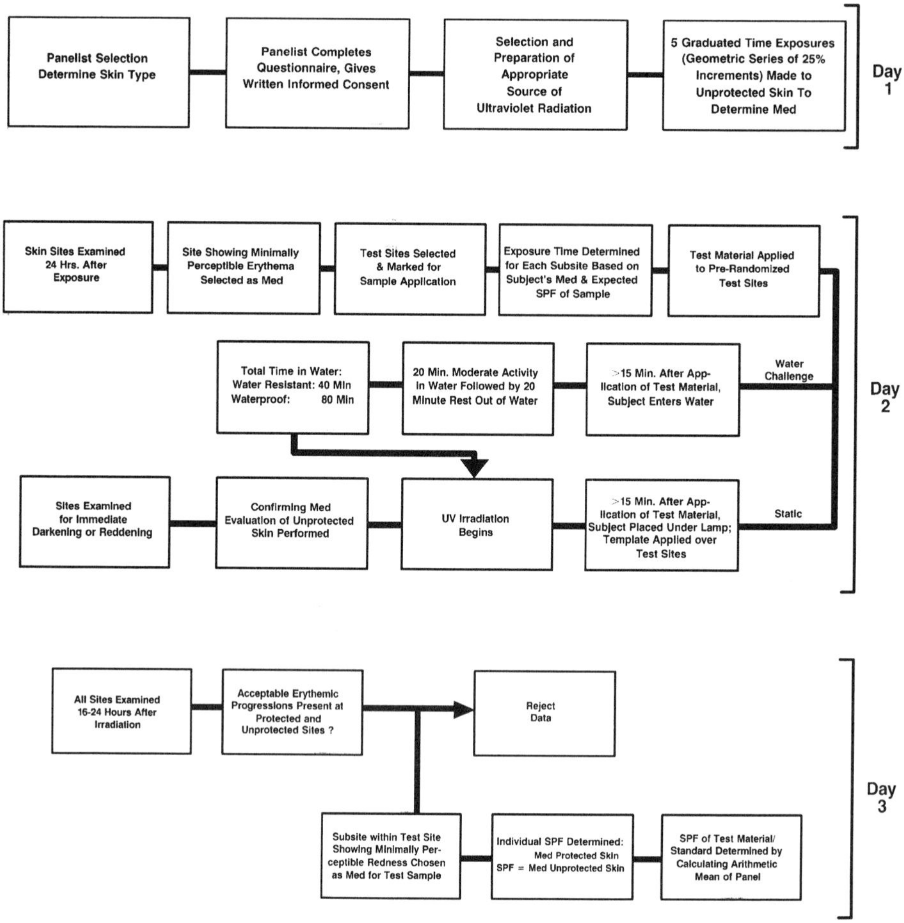

FIGURE 10.2 Sunscreen procedure.

to sunlight such as phototoxic or photoallergic reactions. Individuals with skin diseases such as eczema or psoriasis or who are predisposed to skin cancer should not participate. Likewise, individuals who are sensitive to sunscreens or who are taking any medication that might produce an abnormal response should not be chosen for the panel.

Another screening factor is the skin type of the individual. There are six basic sun-reactive skin types, as shown in Table 10.4 [28]. These skin types reflect the individual's tendency to sunburn and subsequent ability to pigment. Each potential panelist should be observed and questioned about personal sunburn and tanning history. Only fair-skinned volunteers with skin type I, II, or III should be selected. In other words, the individual should burn easily or moderately after 30-60 min of midday sun exposure when the skin has not been previously conditioned to such exposure, such as immediately after winter. Sayre et al. demonstrated that acclimatization occurs in the skin between winter and summer; consequently making an individual respond to sunlight like a less sensitive skin type [29]. Therefore, to be accurate in assessing skin type, the volunteer's response should be based upon recollection of his/her skin reactivity after a prolonged period of abstention from sun exposure.

Finally, the skin on the back of each potential panelist should be examined for sunburn, suntan, scars, or other conditions that might influence the test results.

Once the panelists have been selected, each subject must be informed about the nature of the procedure and any potential hazards, and must provide written consent to participate in the test.

2. Selecting the Irradiation Source

The next step in preparing for the test is choosing the irradiation source. The FDA allows the use of either natural sunlight or a solar simulator. Numerous investigators have reported using natural sunlight to evaluate the efficacy of sunscreen products [3,30-36]. However, while natural sunlight might seem like the best choice, investigators have generally encountered numerous difficulties. In particular, the intensity of ultraviolet radiation in natural sunlight varies. Time of day, season, altitude, latitude, cloudiness, and air pollution can all affect the amount of ultraviolet light reaching the subject. And since it would be difficult or impossible to control these factors, the preferred method of irradiation for sunscreen tests is a solar simulator.

TABLE 10.4 Skin Types

Skin type	Sunburn and tanning history[a]	Example
I	Always burns easily; never tans	People most often with fair skin, blue eyes, freckles; unexposed skin is white[b]
II	Always burns easily; tans minimally	People most often with fair skin, red or blond hair, blue, hazel, or even brown eyes; unexposed skin is white
III	Burns moderately; tans gradually and uniformly (light brown)	Normal average white person; unexposed skin is white
IV	Burns minimally; always tans well (moderate brown)	People with white or light brown skin, dark brown hair, dark eyes (e.g., Mediterraneans, Mongoloids, Orientals, Hispanics); unexposed skin is white or light brown
V	Rarely burns, tans profusely (dark brown)	Brown-skinned persons (e.g., Amerindians, East Indians, Hispanics); unexposed skin is brown
VI	Never burns; deeply pigmented (black)	Blacks (e.g., African and American Negros, Australian and South Indian aborigines); unexposed skin is black

[a]Based on first 30-60 min sun exposure after winter season of no sun exposure.

[b]These individuals may be of Celtic descent (Irish or Scottish); others may have dark hair or brown eyes.

A solar simulator can best be defined as any light source which produces a spectrum of light which closely matches that produced by the sun. In the United States, the preferred system is the xenon arc solar simulator designed by Berger [37]. This system satisfies the requirements stipulated by the FDA:

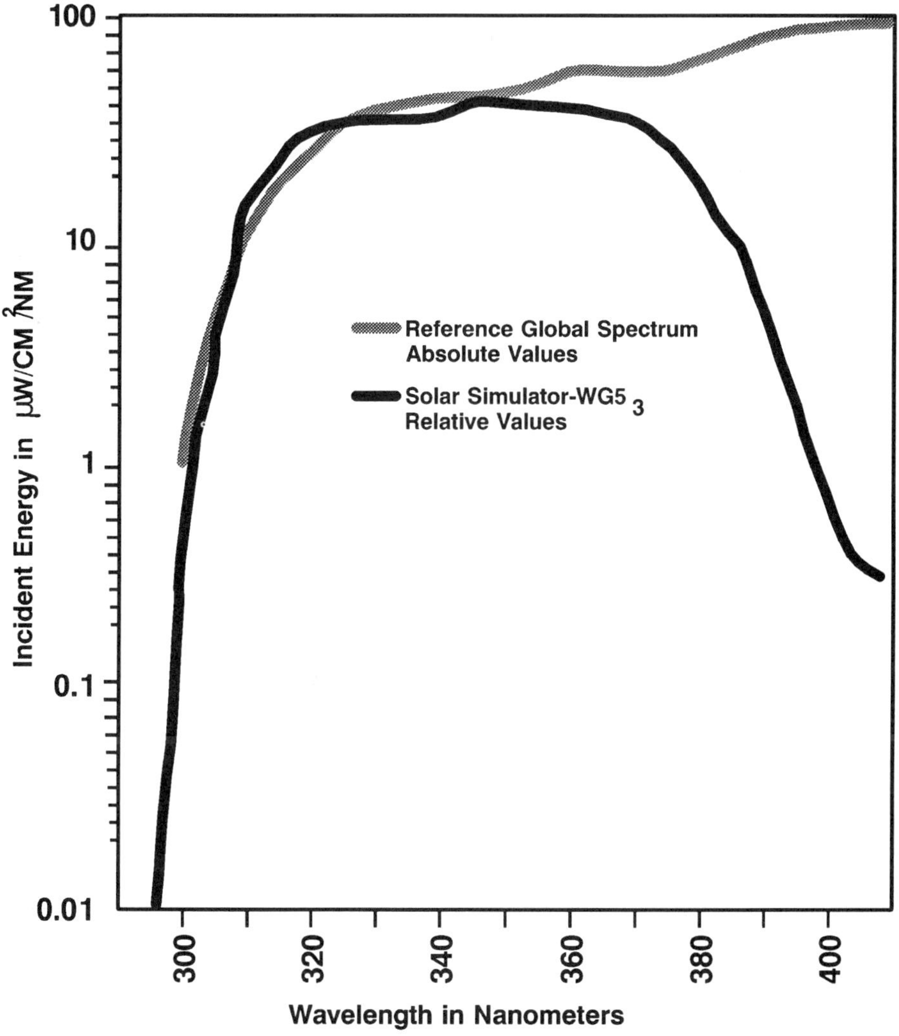

FIGURE 10.3 Solar UV simulator spectrum and spectrum of sun plus skylight. (From Ref. 37, used with permission.)

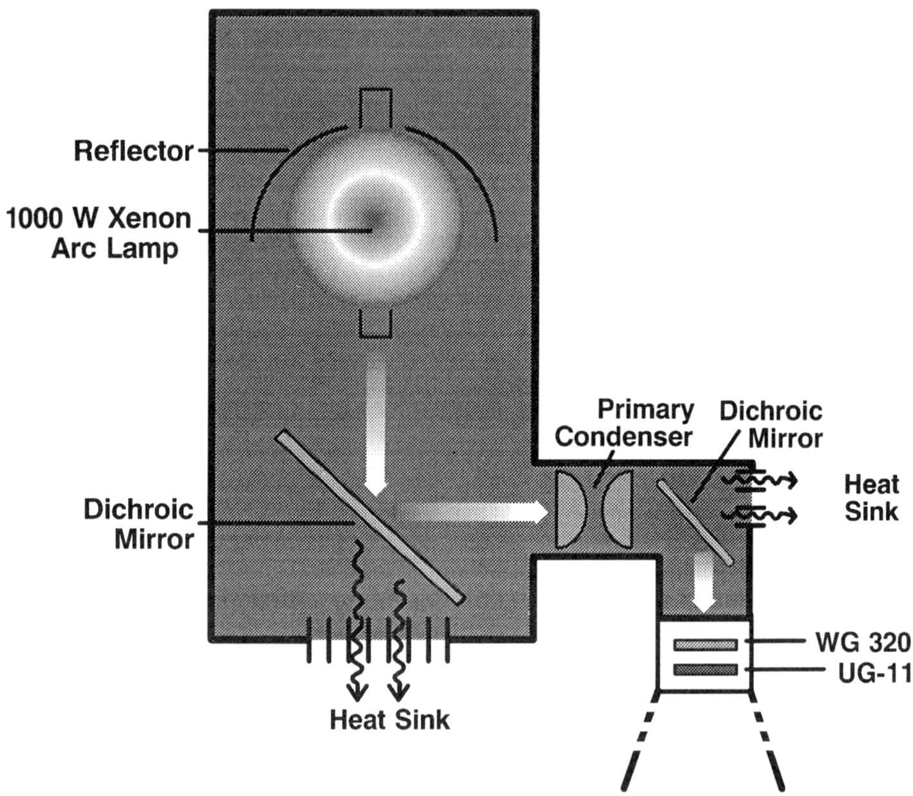

FIGURE 10.4 Configuration of solar simulator.

It emits a continuous UVB spectrum (290-320 nm) with less than 1% of the total energy in short-wave ultraviolet light (shorter than 290 nm). In addition, no more than 5% of the energy it produces to cause sunburn comes from short-wave radiation. Figure 10.3 compares the spectra of natural sunlight and a xenon arc solar simulator.

In a typical xenon arc solar simulator (shown in Fig. 10.4), the xenon arc lamp emits light onto a dichroic mirror mounted at a 45° angle. This mirror reflects the ultraviolet light and transmits about 90% of the visible and infrared light as heat in a heat sink behind the mirror.

The reflected ultraviolet light then passes through a lens that collimates the beam, thus making the intensity fairly uniform across the beam. In some units, especially those with lamps of 1000 W or higher, a second filter such as another dichroic mirror or a water filter may be needed to remove even more of the visible and infrared light that was not removed by the first mirror.

Finally, the beam of ultraviolet light passes through two more filters: a 1-mm thick WG320 filter that removes the short-wave radiation, and a UG-11 filter that absorbs much of the remaining visible or infrared light.

The beam produced by the xenon arc solar simulator is a concentrated source of UVA and UVB light. Although the proportions may vary with the wattage of the system and the quality of its optics, the resulting beam usually contains approximately 30% visible light, 46% UVA radiation, and 24% UVB radiation (the portion of the spectrum that causes sunburn). This is almost 12 times the amount of ultraviolet light found in natural sunlight, yet with a wavelength range that closely matches that of natural sunlight.

The intensity and diameter of the beam vary among different solar simulators. For example, Berger's original design used a 150-W lamp and produced a beam about 1 cm wide. The 1000-W system shown in Fig. 10.4 produces a beam of uniform intensity (within ±10-15%) about 10 cm wide. A small beam can irradiate only one test area at a time, whereas a wider beam lets the investigator irradiate several test areas simultaneously. Berger recently designed a multiport system that more efficiently collects the light emitted from a 150-W lamp and directs it through six liquid light guides, producing six 1-cm wide beams simultaneously.

Regardless of the manufacturer, the energy emitted by the solar simulator should be constant and relatively uniform across the beam. Several factors can influence the output of a solar simulator:

Voltage supply to lamp
Operating temperatures
Lamp age (cumulative hours of use)
Accumulation of dust on optical components
"Solarization" of the lamp envelope and optical components such as filters
Improper alignment of lamp and optical components

With the first four factors, lamp output may vary irradically, making it difficult to achieve beam uniformity and maintain constant intensity. All wavelengths emitted from the lamp tend to be affected equally and output generally decreases. "Solarization" of the lamp envelope or filters occurs with age and changes their transmission characteristics so that transmission of shorter ultraviolet wavelengths is reduced more than the longer wavelengths. Improper alignment yields areas of high intensity (called hot spots) in the beam, which results in nonuniformity of exposure at the skin site.

Regular maintenance and monitoring of the solar simulator becomes very important in producing reliable, reproducible results. Total irradiance of the solar simulator and uniformity of irradiance within the beam can be measured quite adequately on a daily basis using a radiometer equipped with a calibrated cosine-corrected, vacuum or UV-enhanced silicone photodiode detector suitable for measuring ultraviolet radiation. An alternative and widely used device for measuring ultraviolet radiation is the Robertson-Berger (R-B) Meter or Sunburn UV Meter, which gives a reasonable indication of the skin sunburning effect of any light source [38]. Recently, Berger has developed the Erythema UV Intensity Meter which is reported to be more accurate than the Sunburn UV Meter in indicating the sunburning effect of radiation with wavelengths as short as 270 nm.

Evaluation of the emission spectrum should be conducted whenever major changes in the solar simulator occur; e.g., lamp aging, extended use of filters, or change in lamp type or source. To determine the emission spectrum of the solar simulator, a spectroradiometer consisting of a double monochromator, radiometer, and detector is recommended. This equipment is very expensive and consequently may not be readily available to the investigator. An alternative would be to change lamps and filters on a regular basis so that the effects of aging and solarization are minimized, and to use a reliable supplier of such items.

3. Determining the MED of the Test Subjects

Once the panel of test subjects has been selected and the irradiation source chosen, the actual test can begin.

The first step of the process is determining the minimal erythema dose (MED) of each panelist. As stated previously, the MED is the smallest dose or shortest exposure time to ultraviolet light that produces the smallest perceptible redness within about 24 hr.

To determine the MED, each panelist receives five timed exposures on the untreated, unprotected skin of the back. The exposure sites are usually at least 1 cm wide if a solar simulator is used, or 10 cm wide if the irradiation source is natural sunlight. Each exposure lasts 25% longer than the previous one. For example, if the first site receives 38.4 sec of exposure, the second would receive 48 sec, the third would receive 60 sec, and so on.

About 24 hr after these exposures have been completed, the test sites on all panelists are examined for erythema. Ideally, there should be at least one site that shows no erythema and a progression of greater and greater erythema at the sites with longer exposure times. The site with the smallest perceptible erythema (faint redness with distinct edges) is selected as the MED for that individual. This is referred to as the preliminary MED.

4. Testing Sunscreen Efficacy, Including Water Challenge

Once the preliminary MED for unprotected skin has been established for each panelist, the MED for sunscreen-protected skin can be determined.

As with the previous procedure, the skin of the back is usually used for this portion of the test. If a solar simulator is used for irradiation, the test sites should be located between the waist and the shoulder blades, parallel with the waist. Each site should cover about 50 cm^2 of skin. If natural sunlight is used, the entire back from the top of the shoulders to the waist is generally used. In this case, the size of each test site depends on the number of formulations being tested, but the usual size is again about 50 cm^2. The test sites are usually outlined by grids or templates of opaque materials such as medical tape.

Once the test sites have been marked, both the test material and a control formula with a known SPF (8% homomenthyl salicylate, or HMS) can be applied on the prerandomized test sites.

(The maximum number of products that can be tested at one time, including the control, is generally five.)

Two milligrams or 2 μL of sunscreen for each 1 cm^2 of skin is carefully spread to ensure complete, uniform coverage at the test site. The skin is then allowed to dry at least 15 min before the next step begins.

If the manufacturer wants to claim that the product resists being washed off, such a claim can be tested at this point by immersing the test sites in water. Each panelist enters an indoor swimming pool or whirlpool equipped to circulate or agitate the water. The panelists must keep the test sites submerged at all times while performing moderate activity. (This term has never been clearly defined by the FDA; in most tests, the panelist swims several times across the pool or walks about in the water. In a spa, the panelist simply sits in one place while the water circulates to simulate moderate activity.)

Each immersion should last 20 min, followed by a 20-min rest period. The number of immersions necessary depends on the kind of claim that the manufacturer wishes to make. If the manufacturer wishes to claim that the product is water resistant, two immersions are required; if the product is supposed to be waterproof, four immersions must be completed.

Since the HMS control formula is not water resistant, it is not used in this part of the test. Unfortunately, no alternative control formulation has yet been proposed for monitoring the adequacy of the immersion procedure.

After the immersions (or shortly after the application of the sunscreen if no immersions are required), the test sites are irradiated. If a solar simulator is used, five areas within each 50-cm^2 test site are irradiated. The size of each of these five test areas should be at least 1 cm^2. As with the procedure for determining the MED, the length of exposure for each area increases by 25% with each subsequent exposure. The range of length of these exposures is determined by multiplying each panelist's personal preliminary MED by the expected SPF of the product. For example, if a panelist's preliminary MED using a solar simulator is 30 sec of exposure and the sunscreen being tested claims to offer an SPF of 6, then the panelist should be able to withstand 180 sec of exposure before displaying erythema. In this case, the range of exposure times would include one site exposed for 115.2 sec (to produce at least one site with no erythema), one for 144, one for 180, and so on.

If natural sunlight is used, the entire 50-cm^2 test site is irradiated, and at least four such test sites must be used to provide the same incremental exposures for each formula tested.

Before the panelists are released for the day, each individual's MED is again tested at an untreated, unprotected site to confirm the original results and to calculate the SPF values of the HMS standard and the sunscreens being tested. This is referred to as the confirmed MED.

About 16-24 hr after irradiation, each test site is examined for erythema. As with the initial MED determinations, there should be a progression of responses. At least one test site should show no reaction, and the others should show progressively more erythema with increasing exposure. (Of course, if no such progression occurs, the data for that site must be rejected.)

B. Interpreting the Test Results

To determine the SPF of the sunscreen being tested, the MED of the treated area and the MED of the most recent test on unprotected skin (confirmed MED) must be obtained for each test site. The SPF ratings for each panelist are then calculated by dividing the MED of the treated area by the confirmed MED of the untreated area.

The overall SPF of the product is determined by calculating the arithmetic mean of the individual SPF ratings of the entire panel. (The standard error should not exceed ±5% of the mean established in a panel of at least 20 subjects.)

A number of factors can influence the results of this test. Generally, reliable results are obtained when a solar simulator is used. Unfortunately, the same can not be said for tests conducted with natural sunlight. These tests frequently require exposure times of up to 6 hr, especially for products claiming high SPF ratings. Varying environmental conditions over such an extended period often interfere with the amount of ultraviolet light received by the panelists. As a result, the doses are often too small to produce usable data.

Another problem with using natural sunlight for irradiation is that the lengthy exposures required to create an MED also often create a tan. This tan can mask the erythema, making it difficult to determine the MED as well as any progression among the test sites.

Assessment of the MED is a very critical part of the test. Individuals responsible for evaluating test sites for MEDs must not be color blind, should have a clear understanding of the definition of minimal erythema, and, most importantly, be consistent in their scoring process. Also, assessment of erythema should always be conducted in the same location for a given panel. Ideally, an area of the laboratory should be set aside where controlled lighting conditions are available which simulate natural daylight.

These factors should be taken into consideration when determining the SPF and interpreting the results of the tests.

V. CURRENT INTERNATIONAL PERSPECTIVE

The procedure described here for evaluating sunscreens is the method principally used in the United States. And although significant efforts have been made to standardize these procedures worldwide, there are still important differences in the methods used by various countries [39].

For example, there are numerous differences between American and German procedures. In the United States, the irradiation source is the filtered xenon arc lamp; in Germany filtered mercury vapor lamps are used. The spectral distributions of these two light sources are very dissimilar [40,41]. German investigators also use a lower application rate, which may significantly affect results. In Germany, the doses of ultraviolet radiation are administered in 40% increments rather than the 25% increments used here. While these represent the major methodological differences between the two, additional differences do exist and are listed in Table 10.5.

As a result, the SPF ratings for products tested under both procedures have often been reported to be higher in the United States than in Germany [23], raising concern within the private German standard organization, Deutsches Institut fur Normung (DIN). Differences in product application density alone or a combination of the factors discussed above are most likely responsible for the disparity in the test results [40-42].

Australian test methods, on the other hand, are very similar to those of the United States. The few relatively small differences are shown in Table 10.6. These slight differences should produce no significant variations in SPF ratings for equivalent products.

VI. EVALUATION OF UVA PROTECTION

With increasing awareness of the potential for UVA radiation to cause damage to the skin, particularly sun-induced premature aging of the skin, in recent years there has been increased interest in the development of UVA sunscreens. Along with this

TABLE 10.5 Comparison of American and German Sunscreen Testing Standards

	US (FDA)[a]	German (DIN)[b]
Light source	Filtered xenon arc (continuous emission spectrum in UVB region)	Osram Vitalux (300 W) (mercury vapor line spectrum)
	Sunlight	Sunlight testing not addressed
Application rate	2 mg/cm^2	1.50 ± 0.15 mg/cm^2
UV dose increments	25%	40%
Calculation of SPF	Arithmetic mean	Geometric mean
Standard reference	8% homosalate SPF 4.1 ± 0.8	2.7% p-Methoxy-2-ethylhexyl cinnamate SPF 3.7 ± 0.3

[a]Proposed OTC standard method 8/25/78.

[b]Approved standard method DIN norm 67501 (1985).

TABLE 10.6 Comparison of American and Australian Sunscreen Testing Standards

	US (FDA)[a]	Australian (SAA)[b]
Light source	Filtered xenon arc (continuous emission spectrum in UVB region)	Filtered xenon arc (continuous emission spectrum in UVB and UVA regions)
	Natural sunlight	Natural sunlight too variable and unpredictable
UV dose increments	25%	26%
Test panel	At least 20 subjects	At least 10 subjects
Data evaluation	Standard error of mean not to exceed ±5%	Standard error of mean not to exceed ±10%

[a]Proposed OTC standard method 8/25/78

[b]Approved Australian standard AS2604-1986.

interest has come a demand for methods of assessing UVA protection provided by broad-spectrum sunscreens. Most of the methods proposed to date have been based loosely on the existing procedure for evaluating UVB suncreeens for SPF.

Using the SPF method and simply substituting UVA radiation for UVB is impractical, since extremely large doses of UVA (20-50 J/cm^2 for skin types I-III) are required to produce erythema. These doses are approximately 1000 times greater than the dose of UVB required to elicit the same response [43,44]. Such doses imply the use of either extremely long exposure times in normal skin or very intense sources of UVA radiation. For example, on a clear summer day with the sun directly overhead, a typical solar UVB MED for an unprotected, fair-skinned individual would be approximately 16 min; whereas the same individual would require approximately 1.6 hr exposure to solar UVA alone in order to produce an MED (about 30 J/cm^2) [43].

To circumvent the long exposures necessary for evaluating UVA photoprotection in normal human skin, some investigators have chosen to increase the UVA sensitivity of the skin with psoralens or other phototoxic substances [45-51]. While the methods for evaluating topical UVA absorbers have not been standardized, the experimental design tends to be similar from one method to the next. In general, the skin of human volunteers is first sensitized using either oral administration of the phototoxic agent (usually 8-methoxypsoralen) or topical application. After the appropriate waiting period, the sunscreen or vehicle control is applied to the skin at the rate of 2 $\mu L/cm^2$ and spread uniformly over the test site. Fifteen to thirty minutes later, all test sites (treated and untreated) are irradiated with UVA. As in the SPF test, the filtered xenon arc solar simulator appears to be the light source most frequently used although others have been examined [46,48-50]. Exposures within a test site are usually given in 25% dose increments to small subsites and range from 0.1 to 10.0 J/cm^2 in the topical model or 1.0-24.0 J/cm^2 in the systemic model [48-50]. Subjects are examined for erythema between 24 and 72 hr after UVA exposure. Other endpoints such as immediate pigment darkening or delayed melanogenesis have also been used.

Since the protection factors depend on the relationship between the action spectrum of the measured endpoint, the absorption characteristics of the sunscreen agent, and the emission spectrum of the UV source, reported protection factors are true only for the light source used in the evaluation and are not necessarily applicable to normal, nonsensitized individuals.

Misleading protection factors may, in fact, be obtained if the action spectra of the sensitized skin and normal skin are not similar [52].

REFERENCES

1. Shaath, N. A., The chemistry of sunscreens, Cosmet. Toilet., 101(3):55-70, 1986.
2. Knox, J. M., Guin, J., and Cockerell, E. G., Benzophenones: Ultraviolet light absorbing agents, J. Invest. Dermatol., 29:435-443, 1957.
3. Greiter, F., Bilek, P., and Doskoozil, S., History of sunscreens and the rationale for their use. In Principles of Cosmetics for the Dermatologist (P. Frost and S. N. Horowitz, eds.), Mosby, St. Louis, 1982, pp. 187-212.
4. Giese, A. C., and Wells, J. M., Sweat-and-water-resistant sunburn preparations, J. Am. Pharm. Assoc., 35: 208-212, 1946.
5. Pathak, M. A., Fitzpatrick, T. B., and Frenk, E., Evaluation of topical agents that prevent sunburn—Superiority of para-aminobenzoic acid and its ester in alcohol, N. Engl. J. Med., 280:1459-1463, 1969.
6. Langner, A., and Kligman, A. M., Further sunscreen studies of aminobenzoic acid, Arch. Dermatol., 105:851-855, 1972.
7. Fitzpatrick, T. B., Pathak, M. A., and Parrish, J. A., Protection of human skin against the effects of the sunburn ultraviolet (290-320 nm). In Sunlight and Man (T. B. Fitzpatrick, M. A. Pathak, L. C. Harber, M. Seiji, and A. Kukita, eds.), University of Tokyo Press, Tokyo, 1974, pp. 751-765.
8. Willis, I., and Kligman, A. M., Aminobenzoic acid and its esters, Arch. Dermatol., 102:405-417, 1970.
9. Macleod, T. M., and Frain-Bell, W., A study of chemical light screening agents, Br. J. Dermatol., 92:417-425, 1975.
10. Kraft, E. R., Hoch, S. G., Quisno, R. A., and Newcomb, E. A., The importance of the vehicle in formulating sunscreen and tanning preparations, J. Soc. Cosmet. Chem., 23:383-391, 1972.
11. Brown, H., Sunscreens need vehicles, Cosmet. Toilet., 101(3):51-54, 1986.

12. Food and Drug Administration, Sunscreen drug products for over-the-counter human use, Fed. Reg., 43(166): 38206-38269, August 25, 1978.

13. Eiermann, H. J., The regulation of suntan and sunscreen products, Cosmet. Toilet., 98(3):73-78, 1983.

14. Mauer, T., Photocontact allergenicity. In Contact and Photocontact Allergens—A Manual of Predictive Test Methods, Marcel Dekker, New York, pp. 111-123, 1983.

15. Smith, J. G., Jr., Davidson, E. A., Sams, W., Jr., and Clark, R. D., Alterations in human dermal connective tissue with age and chronic sun damage, J. Invest. Dermatol., 39:347-350, 1962.

16. Kligman, A. M., Early destructive effects of sunlight on human skin, J.A.M.A., 210:2377-2380, 1969.

17. Lauker, R. M., Structural alterations in exposed and unexposed aged skin, J. Invest. Dermatol., 73:59-66, 1979.

18. Gilchrest, B. A., Soter, N. A., Stoff, J. S., and Mihm, M. C., jr., The human sunburn reaction: Histologic and biochemical studies, J. Am. Acad. Dermatol., 5:411-422, 1981.

19. Kligman, L. H., Protective effect of sunscreens against photoaging, Drug Cosmet. Ind., 137:30-34, 1985.

20. Robins, P., The need and the challenge, Skin Cancer Found., 3:1, 1985.

21. Lew, R. A., Sober, A. J., Cook, N., Marvell, R., and Fitzpatrick, T. B., Sun exposure habits in patients with cutaneous melanoma: A case control study, J. Dermatol. Surg. Oncol., 9:981, 1983.

22. Johnson, E. Y., and Lookingbill, D. P., Sunscreen use and sun exposure: Trends in a white population, Arch. Dermatol., 120:727-731, 1984.

23. Henne, W., In-vivo determination of the sunscreen factor of cosmetic preparations, history and the present state-of-the art, Parf. Kosm., 64:415, 1983.

24. Schulze, R., Einige versuche und bemerkungen zum problem der handelsublichen lichtschutzmittel, Parf. Kosm., 37:310-315, 1956.

25. Schulze, R., The efficiency of ultraviolet absorbents and the usual commercial sun screen remedies, Strahlenther, 119:312, 1962.

26. Greiter, F., Sun protection factor—Development, methods, Parf. Kosm., 55:70-75, 1974.

27. Ippen, H., Hoppe, U., Wiskemann, A., and Tronnier, H., Biological evaluation of sun protection agents, Artzliche Kosm., 7:102-106, 1977.

28. Wolff, K., Gschnait, F., Honigsmann, H., Konrad, K., Parrish, J. A., and Fitzpatrick, T. B., Phototesting and dosimetry for photochemo-therapy, Br. J. Dermatol., 96: 110, 1977.

29. Sayre, R. M., Desrochers, D. L., Wilson, C. J., and Marlowe, E., Skin type, minimal erythema dose (MED), and sunlight acclimatization, J. Am. Acad. Dermatol., 5:539-443, 1981.

30. Katz, S. I., Relative effectiveness of selected sunscreens, Arch. Dermatol., 101:466-468, 1970.

31. Fitzpatrick, T. B., Pathak, M. A., and Parrish, J. A., Protection of the human skin against the effects of the sunburn ultraviolet (290-320 nm). In Sunlight and Man (T. B. Fitzpatrick, ed.), University of Tokyo Press, Tokyo, 1974, pp. 751-765.

32. Greiter, F., Doskoczil, S., and Bilek, P., Currently used sunscreen materials—Formulation and testing, Cosmet. Sci., 1:153-170, 1978.

33. Sayre, R. M., Desrochers, D. L., Marlowe, E., and Urbach, F., The correlation of indoor solar simulator and natural sunlight, Arch. Dermatol., 114:1649-1651, 1978.

34. Greiter, F., Bilek, P., Doskoczil, S., Washutil, J., and Wurst, F., Methods for water resistance testing of sun protection products, Int. J. Cosmet. Sci., 1:147-157, 1979.

35. Eaglstein, W. H., Taplin, D., Mertz, P., and Smiles, K. A., An all-day test for the evaluation of a topical sunscreen, Am. Acad. Dermatol., 2:513-519, 1980.

36. Azizi, E., Kuschelevsky, A. P., and Schewach-Millett, M., Efficacy of topical sunscreen preparations on the human skin: Combined indoor-outdoor study, Isr. J. Med. Sci., 20:569-577, 1984.

37. Berger, D., Specification and design of solar ultraviolet simulators, J. Invest. Dermatol., 53(3):192-199, 1969.

38. Berger, D. S., The sunburning ultraviolet meter: Design and performance, Photochem. Photobiol., 24:587-593, 1976.

39. Morton, J. J. P., and Murphy, E. G., Sun protection factors: Regulatory issues and methods of determination. In Photobiology of the Skin and Eye (E. M. Jackson, ed.), Marcel Dekker, New York, 85-103, 1986.

40. Sayre, R. M., and Agin, P. P., Comparison of human sun protection factors to predicted protection factors using different lamp spectra, J. Soc. Cosmet. Chem., 35:439-445, 1984.

41. Kaidbey, K. H., Comparison of sun protection factors obtained with a xenon solar simulator and an Ultravitalux lamp, Photodermatology, 4:154-159, 1987.

42. Kaidbey, K. H., and Kligman, A. M., Laboratory methods for appraising the efficacy of sunscreens, J. Soc. Cosmet. Chem., 29:525-536, 1978.

43. Parrish, J. A., Anderson, R. R., Urbach, F., and Pitts, D., Immediate and short-term effects of ultraviolet radiation on skin. In UVA: Biological Effects of Ultraviolet Radiation with Emphasis on Human Responses to Longwave Ultraviolet, Plenum Press, New York, 1978, pp. 107-139.

44. Whitman, G. B., Leach, E. E., De Leo, V. A., and Harber, L. C., Comparative study of erythema response to UVA radiation in humans and guinea pigs. In The Biological Effects of UVA Radiation (F. Urbach and R. W. Gange, eds.), Praeger, New York, 1986, pp. 79-86.

45. Kaidbey, K. H., and Kligman, A. M., Laboratory methods for appraising the efficacy of sunscreens, J. Soc. Cosmet. Chem., 29:525-536, 1978.

46. Jarratt, M., Hill, M., and Smiles, K., Topical protection against long-wave ultraviolet A, J. Am. Acad. Dermatol., 9:354-360, 1983.

47. Diffey, B. L., and Farr, P. M., An evaluation of sunscreens in patients with broad action-spectrum photosensitivity, Br. J. Dermatol., 112:83-86, 1985.

48. Gange, R. W., Soparkar, A., Matzinger, E., Dromgoole, S. H., Sefton, J., and De Gryse, R., Efficacy of a sunscreen containing butyl methoxydibenzoylmethane against ultraviolet A radiation in photosensitized subjects, J. Am. Acad. Dermatol., 15:494-499, 1986.

49. Holzle, E., and Plewig, G., Evaluation of UVA protective agents in humans. In The Biological Effects of UVA Radiation (F. Urbach and R. W. Grange, eds.), Praeger, New York, 1986, pp. 280-285.

50. Lowe, N. J., Dromgoole, S. H., Sefton, J., Bourget, T., and Weingarten, D., Indoor and outdoor efficacy testing of a broad-spectrum sunscreen against ultraviolet A radiation in psoralen-sensitized subjects, J. Am. Acad. Dermatol., 17:224-230, 1987.

51. Kaidbey, K., and Gange, R. W., Comparison of methods for assessing photoprotection against ultraviolet A in vivo, J. Am. Acad. Dermatol., 16:346-353, 1987.

52. Brown, S., and Diffey, B. L., The effect of applied thickness on sunscreen protection: In vivo and in vitro studies, Photochem. Photobiol., 44:509-513, 1986.

11

Bubble Bath and Bath Products

William C. Waggoner

Tewksbury Research, Lebanon, New Jersey

I. INTRODUCTION

Sitting in a bath tub filled with water can expose the skin and urogenital and anal areas to products sold as bath additives, shampoos, soaps, microbes, urine, feces, and a multitude of other unsuspected potential insults. Although the literature is sparse with reports of human irritation from bath products, it is advisable to the manufacturer of any product associated with the bath to obtain data on the safety of its product for future reference should questions arise. This is especially important for products that would be targeted for children.

A. History

A 1955 report [1] discussed 116 adult patients treated by the author for symptoms of vaginitis, which he attributed to "chemical irritation from soap products used." However, among the sources mentioned in the report as potential irritants were soaps, hexachlorophene, douches, bath salts, and bubble bath. The report does not mention which products could have been attributed to any particular incidence of irritation.

Another report [2] was based entirely on the author's experiences with his own child's symptoms of penile irritation and attributed the problem to bubble bath use. However, the boy's twin brother, who shared the baths, developed no symptoms. This suggests that the child may have been particularly sensitive to the product, or, in fact, the irritation may not have been due to the bubble bath at all.

Two other studies [3,4] are based on the same patients. The articles concern 10 of the author's adult patients, all of whom apparently suffered independently from some medical disability. At least seven of the patients were diabetic, a condition that the author admitted renders a person especially susceptible to urinary tract problems. Moreover, he identifies and discusses numerous products, including scented toilet tissues, soap and water, douches, sprays, and bubble baths, as "potential sources of irritation."

Another article [5] described four cases that were "representatives" of the 16 children in the total study. The writer attributed their urinary tract symptoms to the use of bubble bath products. However, he stated that in each case the child had used bubble bath for between one and "several" years without adverse reaction prior to onset of the symptoms described in the report. Therefore, there is doubt whether bubble baths were the exclusive cause of the symptoms.

B. Complaints

An analysis [6] of bubble bath complaints conducted by the Office of Planning and Evaluation (OPE) of the U.S. Food and Drug Administration (FDA) showed that 165 bubble bath complaints were submitted to the FDA over the 7-year period, 1967-1974. This number includes complaint reports aggressively solicited by the Federal Trade Commission bubble bath media campaign conducted during 1973 in Cleveland, Ohio, Atlanta, Georgia, the states of Illinois, Washington, California, and other areas. From a high of 60 reports in 1973, the consumer complaints about bubble bath reported to the FDA from the entire country steadily declined to 31 in 1974, 19 in 1975, 13 in 1976, 10 in 1977, four in 1978, five in 1979, and five in 1980. In addition, data collected by the FDA through voluntary filing of cosmetic product experiences by cosmetic companies in concert with 21 CFR 730, initially confirmed this trend. Bubble bath experiences were 27 in 1974, 30 in 1975, and 10 in 1976, although there were 37 in 1979, 35 in 1980, 45 in 1981, 38 in 1982, and 116 in 1983.

This rate of complaints of bubble bath products is also documented by data collected by the Consumer Product Safety Commission through its National Electronic Injury Surveillance System (NEISS). From July 1, 1977 to June 30, 1978, these data were analyzed by the FDA, and showed that of 26 cosmetic product categories evaluated for "injuries," including dermatitis, "bath products," and "bubble bath products," were reported to present the lowest incidence of injury of all cosmetic product categories.

C. Problems

FDA records indicated that any problems associated with bath products appeared generally to be related to (1) childrens' products, (2) products marketed in powder rather than in liquid form, (3) products in containers permitting excessive use, and (4) products containing alkylarylsulfonates as the surface-active agent.

The Office of Planning and Evaluation's (of the FDA) analysis of bath product complaints for 1967-1974 shows that 98 of 165 complaints received by the FDA over the 7-year period were attributable to one brand of childrens' bubble bath. However, 93 of the 98 reports concerned the powdered version of the product. An additional 24 complaints named another particular brand of childrens' bubble bath, but did not indicate whether or not the brand was powder. The problems with the first product may have been attributable to direct contact with the undissolved powder in the bath water. The problems with bath products may have been related to excessive use.

Use of excessive amounts of bubble bath, just as the use of excessive amounts of any cleaning product, would be more likely to cause irritation. Some adults and unsupervised children may be tempted to use excessive quantities of the product to produce "even more bubbles." Bath products in individual-use containers lessen this risk.

In 42 FR 5368 (early 1977) the FDA requested that manufacturers of bubble bath reduce the concentration of alkylarylsulfonates in their products. The statistics clearly indicate a decrease in the already-low complaint level following the FDA's request for reduced use of alkylarylsulfonates. From 1973 to 1980 the annual complaint rates for bubble bath products were 60, 31, 19, 13, 10, 4, 5, and 5, and appear to reflect a reflect a reduced use of alkylarylsulfonates.

Thus, with a history of low-level, but evident, problems associated with bath product use, it is prudent to test each bath product on a selected group of individuals. After preclinical vaginal screening in oophorectomized rodents, it has been our practice to clinically use a group of prepuberal females after a satisfactory preclinical test. Following are examples of a protocol, consent form, instruction and appointment sheet, and report forms.

APPENDIX

II PROTOCOL

Protocol No.________

Study No.___________

I. TITLE

An Exaggerated Use Study of Children's Bubble Bath in Prepuberal Females

II. SPONSOR AND INVESTIGATOR

Sponsor: [Name, address, and telephone number of Sponsor]

Investigator: [Name and title of investigator]

III. SITE OF STUDY

[Name, address, and telephone number of study location]

IV. INSTITUTIONAL REVIEW BOARD

Prior to acceptance of investigational materials by the investigator, written approval of this protocol must be received by the Sponsor from an appropriately constituted Institutional Review Board.

V. BACKGROUND AND RATIONALE

A prior history of some early bubble bath formulations revealed a propensity to precipitate vaginal, vulvular, and/or dermatological reactions. Reformulation to milder detergent systems caused these types of reactions to wane in this product group. Animal work with this formula [formula code number] gave data that showed a low probability of inducing reactions under use conditions. This project involves testing this product under controlled, exaggerated conditions to ensure that the product has negligible possibility of causing reactions under normal consumer use. Data from a Cosmetic, Toiletry, and Fragrance Association survey shows that normal bubble bath usage in the upper 90th percentile of the USA is approximately two times per week. Thus, the seven-day usage of 140 minutes per week is an exaggerated use.

VI. OBJECTIVES

This study will evaluate a bubble bath under exaggerated use conditions in prepuberal females.

VII. DESIGN

This is a controlled non crossover, open clinical study related to safety. The study will be controlled by selection of a specific target population for whom the product is intended. Product will be tested in the

APPENDIX 11.1 (continued)

intended use arena in an exaggerated fashion.

VIII. SUBJECT DEFINITION

Twenty-five, healthy prepuberal females, ages 6-13. They must be initially free from any signs of irritation on the areas which would be exposed to the product under use conditions. Subjects may not have colds, diarrhea, or any illness that may alter the test outcome.

IX. TREATMENT DEFINITION

Each evening for seven consecutive evenings, the bath tub will be half filled (18-20 gallons) with tap water. The temperature will be in the comfortable range (90-100°F). After the water has been turned on and adjusted for temperature with the drain closed, three cap ful s of the test product are poured into the tub under the tap so that the force of the tap discharge hits the product and causes bubbles to form. Each subject then climbs into the tub and sits for 20 minutes. The subject then gets out of the tub and dries with a clean soft towel without rinsing. The above routine is done only with parental supervision.

X. CONCURRENT TREATMENT

The test product is the only material to be used during the trial period. Specifically excluded in the tub are soaps, bath oils, fragrances, or any bath additive. Fragrance treated toilet tissue may not be used during this study.

APPENDIX 11.1 (continued)

XI. CLINICAL AND LABORATORY MEASUREMENTS

Within 24 hours prior to the start of the 7-day exposure period and within 24 hours following the end of the exposure period, each subject will undergo a thorough examination of the exposed areas including waist, abdomen, buttocks, anus, vulva, groin, thighs, legs, ankles, and feet. Any sign of irritation will be noted and recorded on the subject report form. All examinations will be done by the investigator in the presence of a clinical technician and at least one parent/guardian.

Three capfuls of bubble bath measure approximately 23 milliliters. Twenty gallons of water translate to 75.6 liters. The three capfuls of test product in a half tub of water gives a final concentration of about .03%.

XII. CRITERIA FOR EFFICACY AND SAFETY

A. EFFICACY

Efficacy is not a criterion of this study.

B. SAFETY

Safety of the product will be evaluated on the assessment of reactions that occur.

XIII. STATISTICAL METHODS

Data will be analyzed by appropriate nonparametric methods from values established on an ordinal scale.

APPENDIX 11.1 (continued)

XIV. ADMINISTRATIVE ASPECTS

A. Material Packaging

The sponsor will provide 30 bottles with at least 250 ml (8+ fluid ounces) per container. The bottle and cap liner materials will be identical to the marketed product. The bottle cap will be equal in volume to the marketed product. Each bottle will be appropriately labeled with use instructions. Bottles will be consecutively numbered from 1 through 30. Bottles will be weighed at the study site prior to distribution to each subject. Bottles will also be weighed after their return to the study site at the end of the study. All unused study materials will be returned to the Sponsor.

1. LABEL

> Subject#__________
> Place 3 capfuls of product under running water in tub, which is to be filled half way (18-20 gallons) with (comfortable temperature) water (90-100°F).
>
> [Address and phone number of Clinical Laboratory]

APPENDIX 11.1 (continued)

B. Report Forms

Forms for this study are:

1. Subject Evaluation Form
2. Subject Assignment Form
3. Informed Consent Form
4. Medical History Form

XV. SPECIAL INSTRUCTIONS

A. Investigator

None

B. Subjects

All subjects and an accompanying parent or guardian will read the Informed Consent Form and both sign prior to preliminary acceptance to the study. They will then complete a Subject History Form. Final acceptance to the study will only be after the Investigator has reviewed the Subject History Form and an uneventful clinical examination. Use instructions for the product will be on each bottle. General Instructions and Appointments Form will be given with the product to each subject/parent.

SIGNATURES

INVESTIGATOR: ____________________

DATE__________ Business Phone__________

SPONSOR REPRESENTATIVE:

____

DATE__________ Business Phone__________

APPENDIX 11.1 (continued)

INFORMED CONSENT STATEMENT

The clinical research evaluation for which you/your daughter volunteered is designed to determine the extent, if any, to which the product could produce an irritation reaction in an exaggerated use situation. Twenty-five female subjects between the ages of 6 and 13 years will be involved in the study, which is scheduled to last seven days. The test product, which has been tested in animals and found to have a low propensity to produce reaction, is classified as cosmetic product and is a detergent-like solution.

The subject will initially undergo a physical examination including waist, abdomen, buttocks, anus, vulva, groin, thighs, legs, ankles, and feet. All examinations will be done by Dr._______ in the presence of a clinical technician and at least one parent/guardian. You will use the bubble bath product (3) capfuls in 18-20 gallons of bathtub water every evening for 7 consecutive days and exactly follow the instructions. Specifically, this will consist of sitting in the tub containing the bubble bath for 20 minutes. The test product is the only material to be used during the trial period. Specifically excluded in the tub are soaps, bath oils, fragrances, or any bath additive. There is to be no washing below the waist during the test period. Fragrance-treated toilet tissue may not be used during this study. After the seventh day you will return to the clinic for another examination.

APPENDIX 11.2

Minimal risks to study participants are anticipated during the study. There may be reactions due to irritation or allergy. However, the chance of an allergy is considered unlikely.

The participants in this study will not benefit from the applications of test materials, but test results may allow a new product to be marketed.

The Clinic will keep confidential information concerning you that is obtained in connection with this experiment, except that the Company whose product is being tested will receive a copy of the study. In addition the Food and Drug Administration and others in certain legal actions may inspect the records of the study.

If, in the course of this experiment you experience illness, discomfort, or injury, which appears to be a result of the experiment, The Clinic will provide you with medical care at no cost to you.

APPENDIX 11.2 (continued)

If you have any questions about this experiment or your rights in the case of side effects or injury, contact any staff member who will arrange for you to meet with the Investigator.

INFORMED CONSENT

For Panelist:

I understand that I will be paid an agreed upon amount for completion of this study. I know that I may drop out at any time. If I drop out on my own accord for personal reasons or am dismissed for refusal to obey rules or follow directions, I will not be paid.

For Parent/Guardian:

I have read all the above information and was given an opportunity to ask questions as a parent or guardian about any part of it. Answers to such questions (if any) were satisfactory. I am eighteen years of age or older and freely and without reservation give my informed consent for my daughter to serve as a panelist in this experiment.

PANELIST SIGNATURE____________________ DATE__________
PARENT OR GUARDIAN SIGNATURE____________________
ADDRESS_______________ CITY______________ ZIP______
TELEPHONE NUMBER_______________

DO NOT WRITE BELOW DOUBLE LINE

PANELIST NUMBER____________
WITNESSED BY ____________________ DATE__________

APPENDIX 11.3

MEDICAL HISTORY

CLINICAL TEST VOLUNTEER

DATE____/____/____

NAME:______________________________AGE______BIRTHDATE____/____/____

ADDRESS:______________________________TOWN:________________STATE:_____

PARENT'S NAME:____________________________________PHONE:______________

RACE: CAUCASIAN [ ] BLACK [ ] HISPANIC [ ] OTHER:__________

ANSWER EACH ITEM OR QUESTION BELOW WITH EITHER A YES OR NO

1. DERMATOLOGICAL DISORDERS:
   ATHLETE'S FOOT [ ] ECZEMA [ ] PSORIASIS [ ] ACNE [ ]
   OTHER [ ]_______________DANDRUFF [ ] DRY SKIN [ ]
   IRRITATIONS OR RASH____

2. ALLERGIES:
   HAYFEVER [ ] ASTHMA [ ] POISON IVY [ ] COSMETICS [ ]
   FOODS [ ] SOAPS [ ] DETERGENTS [ ] METAL, JEWELRY [ ]
   LEATHER, FABRICS, DYES [ ] MEDICATIONS________________
   BATH ADDITIVES, IF SO WHICH ONE(S) ?_____________________

3. OTHER SIGNIFICANT MEDICAL PROBLEMS:

CYSTITIS [ ]	HIGH BLOOD PRESSURE [ ]	RHEUMATIC DISEASE [ ]
DIABETES [ ]	THYROID DISEASE [ ]	LUNG DISEASE [ ]
ULCERS [ ]	KIDNEY TROUBLE [ ]	HEART DISEASE [ ]
CANCER [ ]	LIVER DISEASE [ ]	
VAGINITIS [ ]	GALL BLADDER [ ]	

OTHER: (explain)____

DO YOU NORMALLY: BATHE [ ] SHOWER [ ]
ARE YOU TAKING ANY MEDICATION? IF SO EXPLAIN:____________

DATE____/____/____ REVIEWED BY ____________________

APPENDIX 11.4

17

INSTRUCTIONS AND APPOINTMENTS

Study No.______________

You/your daughter have been accepted for a bubble bath study. She is to bathe each evening for seven consecutive evenings for 20 minutes in a half tub (about 20 gallons) of water (90-100°F) under your supervision using three (3) capfuls from the bottle of the liquid product assigned to you. The test product is the only material to be used during the trial period. Specially excluded in the tub are soaps, bath oils, fragrances, or any bath additives. There is to be no washing below the waist during the test period. Fragrance-treated toilet tissue may not be used during this study. You and your daughter are to return to the clinic within 24 hours of the seventh bathing (see appointment below) for a final examination. You must bring the bottle and its unused contents with you for evaluation. If any questions occur, please call the clinic at [Phone number].

Your next appointment is scheduled for____________________
between 4pm and 7pm.

APPENDIX 11.5

Study No.___________

SUBJECT ASSIGNMENT FORM

Subject#	Initial Bottle Weight(g)	Final Bottle Weight(g)	Differential(g)	Comment

Theoretical Average Usage:
Actual Average Use:

APPENDIX 11.6

SUBJECT EVALUATION FORM

Study No._________

SUBJECT NUMBER_______________

SUBJECT'S INITIALS___________ AGE______

BIRTHDATE____________________ RACE__________

WEIGHT (pounds)______________ HEIGHT (inches)_________

FIRST EXAMINATION

SITE	CLEAR	IRRITATION	COMMENT
WAIST	☐	☐	__________
ABDOMEN	☐	☐	__________
BUTTOCKS	☐	☐	__________
GROIN	☐	☐	__________
ANUS	☐	☐	__________
VULVA	☐	☐	__________
THIGHS	☐	☐	__________
LEGS	☐	☐	__________
ANKLES	☐	☐	__________
FEET	☐	☐	__________

☐ QUALIFIED

☐ NOT QUALIFIED

DATE__________ INVESTIGATOR'S SIGNATURE__________________

APPENDIX 11.7

SUBJECT EVALUATION FORM

Study No.________

SUBJECT NUMBER__________________

SUBJECT'S INITIALS______________

FINAL EXAMINATION

SITE	CLEAR	IRRITATION	COMMENT
WAIST	☐	☐	________________
ABDOMEN	☐	☐	________________
BUTTOCKS	☐	☐	________________
GROIN	☐	☐	________________
ANUS	☐	☐	________________
VULVA	☐	☐	________________
THIGHS	☐	☐	________________
LEGS	☐	☐	________________
ANKLES	☐	☐	________________
FEET	☐	☐	________________

NO ADVERSE REACTION ☐

ADVERSE REACTION ☐ (Fully document below)

____

____

____

____

____

____

DATE________ INVESTIGATOR'S SIGNATURE____________________

APPENDIX 11.7 (continued)

REFERENCES

1. Simmons, R. J., Acute vulvovaginitis caused by soap products, Obset. Gynecol., 6:447-448, 1955.
2. Marshall, S., The effect of bubble bath on the urinary tract, J. Urol., 93:112, 1965.
3. Robert, H. J., Bubble bath cystitis, J.A.M.A., 201:207-208, 1967.
4. Robert, H. J., Bubble bath cystitis and cosmetic vulvitis, neglected hazards, J. Fla. Med. Assoc., 60(8):31-35, 1973.
5. Bass, H. N., "Bubble bath" as an irritant to the urinary tract of children, Clin. Pediatr., 7:174, 1968.
6. Food and Drug Administration, Office of Planning and Evaluation, Consumer Safety Statistics Staff, OPE Study 17, Analysis of Bubble Bath Product Complaints 1967-1974, June, 1975.

12

Cosmetic Mouthwashes

Lewis P. Cancro*

Lever Brothers Company, Edgewater, New Jersey

I. INTRODUCTION

Like Virginia Slims cigarettes, the mouthwash has come a long way. Once considered to be as effective as flavored water, the mouthwash in addition to providing benefits of a desirable cosmetic nature, is now widely recognized as an extremely efficient vehicle for the delivery of a number of therapeutic agents. Mouthwashes have an established performance record for fluoride delivery [1-3] as well as for the delivery of antiplaque and anticalculus agents. Three over-the-counter (OTC) fluoride rinses containing sodium fluoride and 10 prescription (Rx) rinses are available to the consumer [4]. The anticaries effectiveness of these rinses is supported by clinical studies that demonstrate a rinse can provide an anticaries benefit beyond that derived from using a fluoride dentifrice alone. A National Institute of Dental Research Study has tested fluoride rinses in school programs in a regimen of daily and weekly application [5]. The benefit observed is a further endorsement of fluoride rinses.

Mouthwashes containing antibacterial agents—cetylpyridinium chloride, domiphen bromide, benzalkonium chloride, thymol-

---

*Current affiliation: Chesebrough-Pond's, Inc., Trumbull, Connecticut eucalyptol, or conventional flavoring components menthol, mints, etc.—have historically straddled the fence, embracing at times both drug and nondrug claims in their marketing strategies. Claims and/or concepts range from cosmetic to therapeutic:

1. Refreshes the breath
2. Combats morning mouth
3. Kills the germs that cause morning mouth
4. Reduces/controls dental plaque
5. Helps prevent the build-up of supragingival calculus (tartar) formation
6. Provides a gingival health benefit

II. OTC REVIEW OF MOUTHWASHES

In 1972, the U.S. Food and Drug Administration's (FDA) review of 60,000 OTC products commenced. The Oral Care Panel, instituted to review one of 23 OTC drug categories, identified seven pharmacological actions associated with ingredients incorporated in OTC products falling within the scope of their review:

1, Anesthetics/analgesics
2. Antimicrobials
3. Astringents
4. Debriding agents
5. Decongestants
6. Demulcents
7. Expectorants

The antimicrobial mouthrinses, of traditional antiseptic/germ kill posture, were reviewed for their pharmacological action in the context of remedial treatment of a disorder. Although these rinses exert a prophylactic action, such as plaque control to prevent the symptoms of a diseased state of the mouth and/or gingiva, the expert panel recommended that medicinal claims for disease prevention not be allowed. The Oral Care Panel reviewed ingredients capable of exerting some degree of dental plaque control, the primary etiological factor associated with gingival inflammation, but they found dental plaque control as an indication for antimicrobial mouthwash use to be inappropriate. The Oral Care Panel, in effect, agreed with the viewpoint expressed by the National Academy of Sciences-National Research Council, Drug Efficacy Study Group (DESI) [6].

"The Panel has reviewed and agrees with the findings of the National Academy of Sciences-National Research Council, Drug Efficacy Study Group (DESI), which pertain to uses of mouthwashes and gargle preparations. In publishing a proposed statement of policy in the *Federal Register* of August 4, 1970 (35FR12411), the Commissioner of Food and Drugs stated that there is a lack of substantial evidence that those preparations are effective for any of their labeled claims which relate to antimicrobial, antiseptic, germicidal, and analgesic uses.

In addition to mouthwash and gargle preparations for which New Drug Applications are in effect, there are many similar products on the market. The Food and Drug Administration has surveyed the labeling of such products, and finds that many of them make direct or implied claims for benefits relating to antimicrobial, antiseptic, germicidal, or analgesic effects. Available information has been reviewed and has not been found to substantiate such claims; i.e.,

> (c) The Administration will not object to labeling of a mouthwash, mouth freshener, or gargle preparation which offers it for such use as an aromatic mouth freshener (provided the product contains aromatic ingredients); as a refreshing mouth rinse; as an aid to daily care of the mouth, and for causing the mouth to feel clean. The label declaration or implication that an ingredient of such an article is active, when this is used to imply that the article has a prophylactic or therapeutic effect, may cause the article to be misbranded. However, an ingredient may continue to be listed on the label if it does in fact contribute to the nonprophylactic and nontherapeutic and usefulness of the article (e.g., wetting agent, foaming agent).

With the publication of Proposed and Tentative Final Monographs by the OTC Dentifrice and Dental Care Drug Products Panel [7] and the OTC Oral Care Panel, dental plaque control appeared by default to have fallen into an unregulated area. Mouthwashes capable of cleaning a tooth's surface to control plaque growth and claiming such action were thought to represent a cosmetic action, particularly if the plaque control claim was linked to cleaning or deodorizing. Mouthwashes exhibiting antimicrobial action through a germicide to control dental plaque build-up for a real or implied health benefit had no OTC Panel proposed indication for use which encompassed plaque control.

In the past 5.5 years, some 7 years after the Oral Care Panel's review of antimicrobial mouthwashes, over 200 clinical studies have been published demonstrating the antiplaque effectiveness of a number of ingredients incorporated into mouthwashes. Prior to that period, chlorhexidine gluconate [8], hexetidine [9], cetylpyridium chloride; domophine bromide [10], benzalthonium chloride [11], and metal salts (copper [12], zinc [13], stannous fluoride [14] among others) when used alone and in combination with germicides had been reported to provide significant plaque control. More recently, reports on mouthwashes containing ingredients with potentially different mechanisms of plaque control (sanguinarine extract [15], sodium benzoate [16], glycerol phosphate [17], and others) appear to offer additional potential as agents capable of controlling plaque growth.

Plaque inhibition through antimicrobial mouthwashes has been measured in the absence of all oral hygiene procedures in periods varying from 18 hr [18] to 28 days [19]. These studies assess plaque regrowth and provide a measure of a mouthrinse's ability to control plaque during its initial and maturing growth phase. The studies are based on the knowledge that in the absence of mechanical disturbance or cleaning agents, oral bacteria will appear on the salivary pellicle that coats teeth to ultimately colonize and aggregate into a mass termed dental plaque. Plaque regrowth at 18 to 48 hr is the measurement of the initial state of this phenomena, and rinsing with an antimicrobial mouthwash under a prescribed regimen can delay or even dramatically reduce bacterial colonization on a tooth's surface.

While most clinical studies undertaken to measure plaque control in the absence of oral hygiene range from 16 to 72 hr, a few studies have evaluated rinses for up to 28 days under similar conditions. When conducted for an extended duration, an effective rinse must be able to provide plaque control which results in the maintenance of oral health in the absence of oral hygiene. Usually, deterioration of the gingival condition will occur in the absence of mechanical cleaning.

Reduction of gingival inflammation and elimination of spontaneous gingival bleeding are some of the benefits to be derived from good plaque control. A change in the gingival tissue from an inflamed to a healthy state usually lags behind plaque removal and control, and thus a 2- to 3-week period represents a period of time in which these clinical changes can be expected to occur.

Plaque control studies in the presence of normal oral hygiene (toothbrushing) can be conducted in trials extending anywhere from 3 to 6 months. These trials, obviously, are dependent upon behavioral and motivational factors in addition to sustained antimicrobial activity of the product. Maintenance of good oral health or oral health improvement should be the objection of such a prolonged trial.

The window of antiplaque effectiveness for germicidal agents in mouthwashes ranges from approximately 7%, a modest reduction, to 70%—an inhibition of significantly greater magnitude. Clinical test design and plaque scoring indices vary in sensitivity depending on the study's objective. Variation in supplemental oral hygiene, treatment regimen, and longevity of the clinical test can result in different interpretation of antimicrobial mouthwash activity.

The American Dental Association (ADA) has to some extent more clearly defined the difference between the therapeutic and cosmetic attributes of mouthwashes intended to control dental plaque. In December 1985, the ADA formulated dental plaque guidelines [20] and established an ADA seal of recognition of safety and effectiveness for products intended to control dental plaque. This event has provided a pathway by which the antimicrobial mouthwashes can be marketed with professional recognition. In the guidelines, the ADA has decreed that antiplaque mouthrinses must, in addition to plaque control, establish that a gingival benefit has been derived from use of the product over a 6-month period.

In 1986, Peridex, an Rx oral rinse containing 0.12% chlorhexidine gluconate received New Drug approval from the Food and Drug Administration's division of Surgical-Dental Products as an effective antimicrobial rinse. The American Dental Association also reviewed the benefit to be derived from use of Peridex and that professional organization's seal of recognition of effectiveness for plaque control with resultant gingival improvement was awarded to the prescription oral rinse on the basis of two independent 6-month clinical studies [21,22]. Plaque control between dental visits was an important aspect of the benefit delivered by this product. The agent chlorhexidine gluconate has also been reported to be effective in promoting the healing of aphthous ulcers [23].

In 1987, Listerine was also awarded the American Dental Association's seal of recognition for plaque control and gingival improvement on the basis of clinical effectiveness demonstrated in two independent 6-month studies [24,25] and one 9-month study [26]. With the ADA seal for plaque control, Listerine has become the second mouthrinse to receive the seal and currently it is the only over-the-counter nonprescription mouthrinse exhibiting that seal.

Thus, following the OTC regulatory review of antimicrobial mouthwashes which did not identify dental plaque as an appropriate indication for use and the publication of the ADA dental plaque guidelines which did, it would appear that plaque control and plaque removal claims can be encompassed by both cosmetic as well as therapeutic mouthwashes. While the regulatory status of such plaque claims is not as yet defined, the professional status has been resolved by the ADA and is highly specific to a demonstration of plaque control and a resultant gingival improvement over a 6-month period. The ultimate regulatory status of plaque claims (drug/nondrug) will undoubtedly depend upon modifiers used in the context of the claims. Plaque control can be achieved through a cleaning and mechanical action to promote cleaning and/or refreshment; plaque control can be delivered through a germicidal mechanism with a demonstration of a health benefit achieved either from a preventative or curative action.

Dental calculus (tartar) is a hard, resistant, mineralized organic deposit that forms on teeth and has recently become the focal point of marketing interests in the dentifrice category. Two anticalculus mouthrinses, Crest tartar and cavity fighting rinse and Colgate tartar control formula, have appeared in the marketplace. Both rinses utilize soluble pyrophosphates as the anticalculus agent. The Colgate tartar control formula also employs Gantrez, a copolymer that has been shown in their tartar control dentifrice formulation to enhance the activity of the pyrophosphate salts. The Colgate tartar control formula has been reported to control the build-up of supragingival tartar by up to 38% on the basis of two clinical studies [27,28].

While the control of supragingival tartar has not been shown to provide a health benefit, tartar does harbor dental plaque, acquires stain, and can only be removed by a trained dental professional. The American Dental Association, a professional society, currently views the control of supragingival calculus (tartar) as a cosmetic benefit. No formal viewpoint on the antitartar benefit of oral products has been expressed by the Food and Drug Administration at this point in time.

III. MOUTHWASH ADVERTISING

The first national advertising of a mouthwash occurred several years prior to 1923 [29]. The product, Listerine, was not marketed as a mouthwash per se, however, but rather as an all-purpose antiseptic. Over the next three decades, the germicidal action of this product founded a number of claims from dandruff control to sore throat prevention. From the campaigns of the 1920s and 1930s, halitosis became synonymous with unpleasant breath, whereas germ kill continued to provide the rationale for product performance. The advertising campaigns on the brand from 1940 through 1960 promoted the product for therapeutic and breath-freshening benefits. Therapeutic claims of cold and sore throat prevention of the mid 1960s eventually led to corrective adversiting in 1978. The disciplinary action of the Federal Trade Commission clearly resulted in stronger emphasis by all brands on cosmetic advertising; i.e., fights bad breath and lasts for hours.

In the mid 1960s through the 1970s, breath freshening and confidence of an extended benefit became the central advertising position of at least two new mouthwashes. Duration of performance, minty smell, breath cleaning without residual medicine breath, and ability to fight strong odors (onions and garlic) provided the creative themes for the advertising campaigns of the mint mouthwashes (Signal and Listermint).

In 1983, the mouthwash with the oldest therapeutic posture featured advertising suggesting a new therapeutic approach: improvement in oral hygiene through plaque control. By mid 1985, a number of mouthrinses were making this claim.

While the evidence is not conclusive, current consumer motivation for mouthwash purchase seems to be almost entirely cosmetic. The consumer expects mouth refreshment and odor control to be the foremost functional properties of mouthwashes. In 1980, U.S. mouthwash users bought over 3 billion ounces, and approximately two of every three adults surveyed used a mouthwash at least once a week [30]. When the mouthwash is used on a daily basis, it is generally used in the morning following toothbrushing, consumption of breakfast, etc., as the last item on one's personal grooming agenda.

Cosmetic mouthwashes and antimicrobial (therapeutic) mouthwashes making cosmetic claims usually substantiate their functional performance through subjective testing. Cosmetic claims vary as does the nature and the degree of technical support required to substantiate them. Some of the expected and delivered benefits of cosmetic mouthwashes are best typified by the following claims:

Refreshes mouth
Leaves your mouth feeling clean
Fights morning mouth
Combats strong mouth odors
Kills germs that cause bad breath
Makes you kissable

Cosmetic claims that do not fall in the area of puffery, i.e., makes you kissable, are generally supported from direct monitoring of the perceived benefit in a target population (mouthwash users) after product use.

Panels composed of 50-200 product users are selected and appropriately balanced with respect to demographics when two or more products are being compared. The panelists are generally provided with a mouthwash, in the presence or absence of a product concept, to be used at home over a period of 1-2 weeks to assure reasonable exposure. Each panelist will rate or judge the appropriateness of verbal or written descriptors, identifying and comparing product attributes either to a control product or to their own expectation of performance.

Intensity factors establish the degree to which the panelist is aware of and in agreement with a specific product attribute, e.g., refreshing flavor. In the paired testing of two rinses, statistical significance can be determined for numerous performance attributes delivered by the rinses. Frequently, however, one or more attributes of a product will create a "halo" effect, i.e., one product can have a dominant attribute that will influence a panelist's judgment of other attributes. For example, a strong or medicinal flavor may lead the user to think the rinse is more effective. The evaluation of the product by the consumer in a home environment, however, generally permits both a measure of the product's acceptability as well as a substantive basis for a specific claim.

Objective measurements of a mouthwash's performance in breath odor and cleaning can be achieved through a number of test methods involving controlled clinical assessment and/or laboratory and instrumental measurements.

Kills the germs that cause bad breath is an example of a claim whose support is derived from the knowledge that a number of oral microorganisms produce putrefying odors from food substrates. Odoriferous fractions can be generated from the bacterial degradation of proteins/peptides, with disulfide linkages or from aromatic amino acids that can provide an indole odor, both commonly identified as contributors to oral malodor. A laboratory demonstration of the effect of a mouthrinse in suppressing odor producing bacteria has been used to support the germ kill claim.

Combats morning mouth and fights strong mouth odor are examples of claims that are generally supported by direct olfactory assessment of breath odor by one or more trained judges. The tests are conducted in a controlled clinical environment. Judges are trained to consider the magnitude of the odor and the nature of the insult (garlic, onions, etc.). The test design considers balancing of the subjects, blinding of the subjects and judges, and appropriate controls to assess shifting of criteria by the judges as well as other variables.

Subjects are selected to participate in the trial if they are willing to forego their morning oral hygiene procedures and exhibit morning mouth, the strong stale or slightly putrescent odor present on awaking. Judges are selected on the basis of their ability to discriminate different odors and to estimate the intensity of a given odor; training and experience are essential ingredients in their ability to consistently identify an odor and judge its magnitude.

Several clinical testing facilities have developed considerable expertise in odor assessment [31-33]. These facilities usually test mouthwash performance against specific or general malodor in groups of subjects ranging from 12 to 30 or more. A single judge or as many as five can be used to assess odor. Odor intensity scales from 0 to 5 or from 0 to 9 are usually used to determine the magnitude of odor. Each judge assesses the breath odor of the panelists prior to a treatment and over a number of points in time, ranging from immediately after product use up to 4 or more hours.

Most mouthwashes suppress extrinsic malodor for several hours before the malodor intensity rises to an objectionable level. Odor control through flavor masking is relatively short lived, perhaps 20 min, and with malodors induced through abstention from oral hygiene or through foods, the breath quickly regains its initial character if the only mode of odor control by the mouthwash is flavor masking. Flavor fatigue and taste longevity are probably associated with retention and clearance of flavor from the oral cavity but not with odor-suppressing effectiveness. Individuals with an acute sense of taste can identify persistent flavors for 20 to 90 min [34]. This does not, however, coincide with a mouthwash's effectiveness, as certain rinses achieve effectiveness through their antimicrobial or cleaning properties long after the flavor of the mouthwash has expired.

Certain laboratories have developed through the years a relationship of the magnitude of an odor to its generally perceived offensiveness with large panels of consumers. Thus, when a mouthwash suppresses mouth odor to a certain level, its degree of effectiveness can be determined and can be projected over a broad population.

Hedonic assessment of an odor, i.e., a judgment of the acceptability of the odor or the degree to which a judge(s) likes the odor, has been employed in "clinical situations" to ascertain the degree to which a breath odor shifts from a negative state of acceptance to a positive state of acceptance under a treatment regimen. Whether the assessment scale is based on an intensity rating or a hedonic rating, comparative performance of mouthwashes can be achieved and the data derived are amenable to a variety of statistical procedures from which confidence levels can be projected. An effective mouthwash usually suppresses extrinsic oral malodor for 3 or more hours and displays its effectiveness across a significant distribution of the subject population tested.

Advertising campaigns featuring claims of a competitive nature occasionally appear in the mouthwash category:

Lasts twice as long
Preferred two to one
Fights strong mouth odor better

These must, by advertising agency and network standards, meet requirements involving direct comparative testing with an appropriate population. The magnitude of the difference between two treatments must achieve statistical confidence at the 95% confidence level and should also be consumer perceptible.

Cosmetic mouthwashes can deliver benefits through food debris removal, another aspect of oral hygiene. A water rinse can reduce the oral microbial count for approximately 20 min. Alcohol and surfactants, usually part of a mouthwash formula, enhance that benefit. Food debris, exfoliated epithelial cells, and bacteria can be removed through vigorous rinsing. Mouthwashing with a volume of 10-15 ml efficiently coats the mucosal lining, and depending on the time and the vigorousness with which one rinses, crevices and areas beneath the tongue can be flushed to remove residual food debris.

The efficiency of debris removal can be objectively measured through tracers, i.e., food dyes incorporated in food and measured in the expectorant. One can also objectively assess the amount of food debris flushed from the oral cavity by collecting expectorant after rinsing, centrifuging, and measuring the packed debris volume. The principle is routinely employed in hematological measurements in determining the packed red cell volume [35].

Mouthwashes have broad consumer appeal but the mouthwash category has experienced a slight decline in overall growth of approximately 5% in the 5-year period from 1980 to 1985. Current per capita consumption of 12-13 oz/year appears to have stabilized. The fastest growing segment of the mouthwash market is the "all other" category. This category features such specialties as the fluoride rinses, special prebrushing and plaque-controlling rinses, and private label mouthwashes positioned at lower price levels. Mouthwashes for the most part, however, are purchased to promote a sense of confidence. Mouthwashes effectively clean the mouth, refresh one's taste, make the breath fresher, and accomplish this benefit in an application time that rarely exceeds 30 sec. The consumers' expectations of mouthwash performance are high and they are seldom disappointed with a product's performance.

REFERENCES

1. Rugg-Gunn, A. J., Holloway, P. J., and Davies, T. G. H. Caries prevention by daily fluoride mouthrinsing: Report of a three-year clinical trial. Br. Dent. J., 135: 353-360, 1973.
2. Britta, F. The caries preventing effect of mouthrinsing with 0.025% sodium fluoride solution in Swedish children. Commun. Dent. Oral Epidemial., 2:58-65, 1974.
3. Finn, S. B., Moller, P., Jamison, H., Regattieri, and Manson-Heng, L. The clinical cariostatic effectiveness of two concentrations of acidulated phosphate-fluoride mouthwash, J.A.D.A., 90:398-402.
4. American Dental Association, Council on Dental Therapeutics: Accepted Dental Therapeutics, 40th ed., American Dental Association, Chicago, 1984, pp. 404-405.
5. Driscoll, W., Swango, P. A., Horowitz, A., and Kingman, A. Caries—Preventive effects of daily and weekly fluoride mouthrinsing in a fluoridated community: Final results after 30 months, J.A.D.A., 105:1010-1013, 1982.

6. Fed. Reg./Vol. 47, No. 101. Tuesday, May 25, 1982. Proposed Rules.

7. Fed. Reg./Vol. 50, No. 189. Monday, September 30, 1985. Proposed Rules.

8. Loe, H., and Schiott, C. R. International Conference on Periodontal Research, J. Periodont. Res., 4:38-39, 1969.

9. Grenby, T. H., and Saldanha, M. G. The antimicrobial activity of modern mouthwashes, Br. Dent. J., 157:239-242, 1984.

10. Sturzenberger, O. P., AND Leonard, G. J. The effect of a mouthwash as an adjunct in toothcleaning, J. Periodont., 40:299-301, 1969.

11. Guffar, A., and Rustogi, K. N. Effects of sodium phytate and benzalthonium chloride rinses on plaque formation in humans, Caries Res., 16:472-474, 1982.

12. Waerhaug, M., Gjermo, P., Rolla, G., and Johansen, J. R. Comparison of the effect of chlorhexidine and $CuSO_4$ on plaque formation and development of gingivitis, J. Clin. Periodont., 11:176-180, 1984.

13. Harrap, G. J., Best, J. S., and Saxton, C. A. Human oral retention of zinc from mouthwashes containing zinc salts and its relevance to dental plaque control, Arch. Oral Biol., 29:87-91, 1984.

14. Leverett, D. H., McHugh, W. D., and Jensen, O. E. The effect of daily mouthrinsing with stannous fluoride on dental plaque and gingivitis—Four-month results, J. Dent. Res., 60:781-784, 1984.

15. Southard, G. L., Boulware, R. T., Walborn, D. R., Groznik, W. J., Thorne, E. E., and Yankell, S. L. Sanguinarine, a new antiplaque agent: Retention and plaque specificity, J.A.D.A., 108:338-341, 1984.

16. Emling, R. C., and Yankell, S. L. First clinical studies of a new prebrushing mouthrinse. Compend. Contin. Ed., 6:636, 1985.

17. Wycoff, S. J., Morris, M. E., and Newbrun, E. The effect of a mouthrinse containing calcium glycerolphosphate on the chemical composition and development of plaque in humans, J. Dent. Res., 59:23-28, 1980.

18. Harrap, J. J. Assessment of the effect of dentifrices on the growth of dental plaque, J. Clin. Periodont., 1:166-174, 1974.

19. Loe, H., and Schiott, C. The effect of mouthrinses and topical application of chlorhexidine on the development of

dental plaque and gingivitis in man, J. Periodont. Res., 5:79-83, 1970.

20. Council on Dental Therapeutics. Guidelines for Acceptance of Chemotherapeutic Products for the Control of Supragingival Dental Plaque and Gingivitis, American Dental Association, 1985.

21. Segreto, V. A., Collins, E. M., Beiswanger, B. B., De La Rosa, M., Isaacs, R. L., Lang, N. P., Mallatt, M. E., and Meckel, A. H. A comparison of mouthrinses containing two concentrations of chlorhexidine, J. Periodont. Res. Suppl., 23-32, 1986.

22. Grossman, E., Reiter, G., Sturzenberger, O. P., De La Rosa, M., Dickinson, T. D., Ferretti, G. A., Ludlam, G. E., and Meckel, A. H. Six-month study of the effects of a chlorhexidine mouthrinse on gingivitis in adults, J. Periodontal Res. Suppl., 33-43, 1986.

23. Addy, M., Carpenter, R., and Roberts, W. R. Management of recurrent aphthous ulceration. A trial of chlorhexidine gluconate gel, Br. Dent. J., 141:118-120, 1976.

24. DePaola, L. G. et al. Chemotherapeutic inhibition of supragingival plaque and gingivitis development. Submitted for publication.

25. Lamster, I. B. et al. The effect of Listerine antiseptic on reduction of existing plaque and gingivitis, Clin. Prev. Dent., 5:12-16, 1983.

26. Gordon, J. M. et al. Efficacy of Listerine antiseptic in inhibiting the development of plaque and gingivitis, J. Clin. Periodont., 12:697-704, 1985.

27. Singh, S., and Harrison, L. The anticalculus effect of a mouthrinse containing soluble pyrophosphate salts and a copolymer. In press.

28. Rustogi, K. N., and Petrone, M. E. Clinical calculus study of a mouthrinse containing soluble pyrophosphate salts and a copolymer. In press.

29. Fox, S. The Mirror Makers, William Morrow, New York, New York, 1984, p. 97.

30. Mouthwash Usage, Simmons Market Research Bureau, Inc., for Lever Brothers Co., 1985.

31. Consumer Product Testing Service, Inc., 20 Pickering Street, Needham, Massachusetts 02192.

32. Clinical Protocol Associates, Inc., 2125 Center Avenue, Fort Lee, New Jersey 07024.

33. A. D. Little, 25 Acorn Part, Cambridge, Massachusetts 02150.

34. Flavor Persistence, TEV 84-0064, Lever Brothers Company, 1984.

35. Bray, W. E., Clinical Laboratory Methods, The C. V. Mosby Company, St. Louis, 1957.

13

Douches

Bernard Misek, Alan D. Kirsch, and Marvin Kaminsky

Beecham Products, Parsippany, New Jersey

I. INTRODUCTION

A vaginal douche is a liquid preparation used to irrigate the vagina for one or more purposes such as cleansing, deodorizing, producing a refreshing effect, reducing the number of pathogenic microorganisms, or relieving minor vaginal irritations. Whether a douche product is categorized as a drug or cosmetic depends upon its ingredients and their concentrations and the claims made for the product. The safety testing program to qualify a douche product for market introduction is the same whether the product is classified as a drug or a cosmetic.

Presently, douches which are classified as over-the-counter (OTC) drugs contain the active ingredient povidone-iodine at concentrations of 0.15-0.30%. Povidone-iodine is the only active ingredient currently accepted as safe and effective for OTC use by the U.S. Food and Drug Administration (FDA) for the relief of minor vaginal irritations.

Cosmetic douches, such as those containing vinegar, water, fragrances, or other substances, provide a nontherapeutic effect. Among the claims made for cosmetic douches are cleansing, deodorizing, and refreshing.

II. SAFETY TESTING

Prior to any human testing, a thorough review of the formulation is made to insure there are no toxicological reasons to abandon the product. It is essential that the potential for local irritation of the vaginal mucous membranes be virtually nonexistent. Individual ingredients are evaluated for safety based on history of use in similar formulas, literature, supplier's safety data, CTFA Cosmetic Ingredient Review and any other available information. Finished formulations are then evaluated based upon historical data on similar formulations. If existing information are determined to be insufficient to substantiate safety, then appropriate animal toxicology is conducted.

Screening douche formulations in animals is an appropriate method for evaluating vaginal irritation potential. The rabbit vagina is routinely used in these studies [1]. Comparison of the findings in vaginal studies in the rabbit with those of the human vagina has demonstrated good agreement [2]. In addition, contact allergenicity testing in guinea pigs may be conducted to provide the toxicologist with information about the contact allergenicity potential of the formulation during the course of repeated exposures under conditions of normal use. From a review of the total information available, the toxicologist provides expert judgment about the safety of the formulation.

Human safety testing of the formulation commences following the toxicological review and the successful completion of any necessary toxicology studies. The program would usually include repeated insult patch testing on 100 subjects and use safety testing on 50-100 subjects. Repeated insult patch testing is conducted to determine whether or not the formulation is capable of affecting human skin under patch test conditions; and if so, to characterize the mechanism of action as primary irritation, skin fatiguing (cumulative irritation), and/or contact sensitization. In these studies, the skin of patch testing participants is subjected to conditions which are substantially more stressful than the conditions under which human skin would encounter the product in real life situations (occlusion, frequent exposure, higher concentrations, etc.)

Conclusions about the primary and cumulative irritating propensities of the test material are arrived at on the basis of the investigator's interpretation of the results. Taken into consideration when making an interpretation are the incidence, frequency, type, and intensity of the skin responses.

TABLE 13.1 Statistical Table for Extrapolating the Results of a Patch Test Study in Various-Sized Panels[a]

Number sensitized	95% Confidence limits	Size of panel (%)		
		50 people	100 people	200 people
0[b]	Upper	100	100	100
	Lower	92.89	96.38	98.17
1[c]	Upper	99.949	99.975	99.987
	Lower	89.34	94.55	97.24
2[c]	Upper	99.51	99.76	99.88
	Lower	86.28	92.96	96.43
3[c]	Upper	98.74	99.38	99.69
	Lower	83.43	91.47	95.68
4[c]	Upper	97.77	98.90	99.45
	Lower	80.75	90.07	94.96

[a]The table defines the upper and lower percentages of a general population which may, with 95% confidence, be expected to tolerate the test material if 0, 1, 2, 3, or 4 panelists were sensitized during the course of the patch test.

Source: Compiled from Mainland, D., Herrera, L., and Sutcliffe, M. I., Binomial Confidence Limits, Tables for Use with Binomial Samples (Tables IX[b] and V[c]), Department of Medical Statistics, New York University, College of Medicine, 1956.

The responses are assigned numerical grades to indicate the intensities and/or character of the adverse effects seen by the investigator. A sample patch test case report form (Fig. 13.1) is appended. With regard to an assessment of the population risk of sensitization resulting from contact with the formulation, the investigator relies on statistical tables (Table 13.1) which define the confidence limits in extrapolating the expected incidence of sensitization in a general population from the incidence encountered in a small test population. The specific number of females participating in use studies is dependent on a careful evaluation of product ingredients. A minor change from currently marketed douche formulations, such as a slight change in fragrance or preservation level, would necessitate a use study consisting of at least 50 subjects; whereas a major change, such as a new fragrance or preservative, would require at least 100 study participants.

A. Subject Selection

Criteria for selection of use study participants consist of the following:

1. A signed informed consent form and willingness to adhere to the study protocol
2. Premenopausal females between the ages of 18 and 45 years
3. Good general health and a regular menstrual cycle
4. Pap smear tests with either class I or II results, as long as no disease is present that may have an effect on the study outcome
5. Previous douching experience on a regular basis

Participants having a history of gynecological disorders or with acute or chronic diseases are excluded. Among the reasons for exclusion are: (1) complaints of vaginitis with or without discharge; skin disease which might be confused with a skin reaction from the test formulation; and (3) the use of medications, either intravaginal or systemic, which might influence or affect the evaluation of the formulation. However, subjects having had a partial hysterectomy would be allowed to participate in the study.

In addition, females presenting with irritation (soreness) or itching due to the use of feminine sprays, cuts, abrasions, chemical burns, or contact dermatitis are excluded from study participation. Pregnant women, nursing mothers, or mothers less than 2 months post partum are rejected.

B. Study Design

A board certified gynecologist performs a routine gynecological examination on each individual, including a visual and speculum examination of the vulva, vaginal vault, and cervix to determine the presence of irritation or if a disorder exists. A Pap smear is taken to aid the findings of the gynecological exam. Results are recorded on the Initial Medical History and Gynecological Examination case report form (Fig. 13.2).

If a subject is qualified for the use study based on meeting the protocol inclusion requirements, she is provided with a supply of douches designated by subject number and day of use. The douching regimen consists of each subject receiving 12 douches to be used three times a week on alternate days during a 4-week period, excluding the days of menses.

In order to be certain the subjects douche properly during the study, the first douche is performed under the supervision of a qualified individual.

Subjects maintain log sheets (Fig. 13.3) during the course of the study, recording the date and time of day each douching is performed and any adverse experiences. Subjects are required to bring the log sheets to the study site on a weekly basis so that the study director is certain the subject is complying with protocol directions.

As soon as possible after the last douche, a final gynecological exam is performed to determine if the douche solution has caused vulvar, vaginal, or cervical irritation. Findings are recorded on the Final History and Gynecological Examination case report form (Fig. 13.4). The participants' log sheets are reviewed by the gynecologist who will also discuss subjective observations with the individual. The gynecologist prepares a report which includes clinical findings, adverse experiences, and the outcome of unusual findings.

REFERENCES

1. Kaminsky, M., and Willigan, D. A., pH and the potential irritancy of douche formulations to the vaginal mucosa of the albino rabbit and rat, Fd. Chem. Toxic, 20:193, 1982.
2. Eckstein, P., Jackson, M. C. N., Millman, N., and Sobrero, A. J., Comparison of vaginal tolerance tests of spermicidal preparations in rabbits and monkeys, J. Reprod. Fert., 20:85, 1969.

APPENDIX

PATCH TEST

SUBJECT NAME (Last, First, Middle)	AGE	SEX		SUBJECT NUMBER

DATE:													
APPL. NO.:	1	2	3	4	5	6	7	8	9	10	Challenge		
	**	**	**	**	**	**	**	**	**	**	**	**	**
LOCATION:													
SCORE:													
LOCATION:													
SCORE:													

SCORING CRITERIA:

0 = No Reaction
1+ = Slight Erythema
2+ = Marked Erythema
3+ = Marked erythema, edema, with or without a few vesicles
4+ = Marked erythema, edema with vesicles and oozing

LOCATION:

1 = Left Upper Arm (LU)
2 = Left Forearm (LF)
3 = Right Upper Arm (RU)
4 = Right Forearm (RF)

Other Adverse Reaction: [ ] Yes [ ] No If YES, Specify: ____________________

RETURN COMPLETED FORMS:	Investigator's Signature	Date Signed (mo/day/yr)
	Investigator's Name	

FIGURE 13.1

DOUCHE TOLERANCE EVALUATION

INITIAL MEDICAL HISTORY & GYNECOLOGICAL EXAMINATION

SUBJECT #:	SUBJECT NAME: (Last, First, Middle)	DATE:

AGE [|] SEX [ ] WT. [| |] HT. [| |] Ft. | Inches

MEDICAL HISTORY

GYNECOLOGY:____

MENSTRUAL CYCLE: Normal [ ] Abnormal [ ] Cycle Length ________Days

Days since onset of last menses: ________

SURGICAL:____

ALLERGIES:____

PAP TEST: Class________

LIST ANY CHRONIC MEDICATIONS BEING TAKEN (Include Oral Contraceptives)

[ ] NONE (or specify)	DOSAGE	FROM	THROUGH	REASON FOR USE	SIDE EFFECTS

GYNECOLOGICAL EXAMINATION

*Indicate Degree of IRRITATION as Follows: 0 = Absent 2+ = Moderate
1+ = Mild 3+ = Severe

SITE	IRRITATION*	DESCRIBE
Vulva		
Vagina		
Cervix		
Other (specify)		

RETURN COMPLETED FORMS:	Investigator's Signature	Date Signed (mo/day/yr)
	Investigator's Name	

FIGURE 13.2

DOUCHE TOLERANCE EVALUATION - **SUBJECT'S LOG SHEET**

SUBJECT #:	SUBJECT NAME: (Last, First, Middle)	DATE:

Please note date and hour each time you douche.
Return this form to clinic after using product as directed.
DO NOT DOUCHE ON DAYS OF MENSES, but record these days below.

STUDY DAY #	**DATE**	**TIME**	**COMMENTS**

RETURN COMPLETED FORMS:	SUBJECT'S SIGNATURE:	DATE:
	INVESTIGATOR'S SIGNATURE:	DATE:

FIGURE 13.3

DOUCHE TOLERANCE EVALUATION

F I N A L HISTORY & GYNECOLOGICAL EXAMINATION

SUBJECT #:	SUBJECT NAME: (Last, First, Middle)	DATE:

LIST ANY CHRONIC MEDICATIONS BEING TAKEN (Include Oral Contraceptives)

☐ NONE (or specify)	DOSAGE	FROM	THROUGH	REASON FOR USE	SIDE EFFECTS

GYNECOLOGICAL EXAMINATION

*Indicate Degree of IRRITATION as Follows: 0 = Absent 2+ = Moderate
1+ = Mild 3+ = Severe

SITE	IRRITATION*	DESCRIBE
Vulva		
Vagina		
Cervix		
Other (specify)		

SIDE EFFECTS, if any ______________________________

____

COMMENTS BY INVESTIGATOR: ______________________________

____

RETURN COMPLETED FORMS:	Investigator's Signature	Date Signed (mo/day/yr)
	Investigator's Name	

FIGURE 13.4

14

Assessment of Antiaging Products

Peter T. Pugliese

Peter T. Pugliese, M.D. and Associates, Bernville, Pennsylvania

I. INTRODUCTION

Aging must be considered a natural process for all material, whether animate or inanimate. How one defines aging depends to a large measure on the scientific discipline of the individual. For our purposes we shall define aging as a biological process associated with certain biochemical and biophysical changes that relate to reduced functional capacity of the individual through time. This reduced capacity is generally directly proportional to the age of the individual. In this definition the word disease is not used, though aging may also be considered a disease and not as a natural process [1-5].

The many theories of aging may be reduced to two major concepts: The castastrophe theory, and the free radical theory. The scientific literature supports the free radical theory, and it is this concept that will be followed in this chapter. It is necessary to provide the reader with a fundamental understanding of the free radical concept as it relates to aging before we can discuss how this concept can be used to measure various parameters of aging.

A. The Free Radical Concept of Aging

Free radicals are chemical species that are characterized by a molecular structure that is not electronically complete. That is, there are unsatisfied electronic shells that could react with other chemical species rather than forming bonds within the molecule itself. Usually these molecules are chemically unstable and highly reactive, so that they tend to react avidly with other molecules, particularly oxygen. Not all free radicals are unstable or highly reactive; however, the biologically important free radicals are very reactive. Free radicals are usually produced by an initiating force such as ultraviolet radiation or as a result of some associated chemical reaction that produces free electrons as a by-product. Biological oxidations are frequent offenders in this category.

A particularly vulnerable and ubiquitous chemical in the body is the unsaturated fatty acid, most often occurring in the form of a phospholipid, and situated in the cell membrane. When a free radical reacts with a fatty acid molecule the resultant free radical fatty acid will frequently react with molecular oxygen, forming a peroxy radical. The resultant peroxy radical will seek other sources of electrons to satisfy its unstable nature and will prey upon adjacent fatty acid molecules. Thus, a chain reaction is produced that is capable of destroying a complete cell membrane or microsome in a fraction of a second.

By-products of the chain reactions have been identified in cross-linking reactions with proteins of many types, which results in decreased function of these proteins [6]. There are many agents produced in the body that are capable of cross-linking reactions. These include aldehydes, sulfur, various quinones, antibodies, polybasic acids, and polyvalent metals besides the free radical-induced reactions. The net result of all these reactions must be viewed as reducing the functional capacity of the organism if they are to be related to the aging process. The two areas that will be discussed in this chapter that relate to the free radical process concern the effect on viscoelastic properties of the skin and the presence of lipid peroxidation products in the skin. A method will be described to measure viscoelastic properties of aging skin and the ability of products to alter these properties. The second topic will describe a method to measure lipid peroxides in the skin as a function of age or ultraviolet exposure and how products may be evaluated by this system.

II. MEASUREMENT OF VISCOELASTIC PROPERTIES OF SKIN

In the aging process there are changes in the connective tissue that are subtle and not easily detected until a secondary manifestation appears. Such is the case in the vascular system where blood vessels are slowly altered owing to elastin degradation or modification. In the skin, early changes in viscoelastic properties are visible as sagging and increased extensibility. These changes are measurable in vivo, and appear to represent a quantitative parameter that can serve as a marker of product efficacy [7].

A. Review of Methods

Currently there are four methods that are being used to measure viscoelastic parameters of the skin [8-11]. These are:

1. Indentation techniques
2. Extension techniques
3. Torsion techniques
4. Suction techniques

A comprehensive review of these methods is not the intent of this chapter, for we shall discuss only one method in detail, and that is the suction technique. I must point out, however, that no method at present is capable of measuring all the mechanical properties of skin satisfactorily. The skin is a living tissue and defies exactly repeatable measurements. Any method used to measure a living system invariably must in some manner perturb that system, and thus distort the natural state. A few comments on the first three techniques will acquaint the reader with the basic principles of these methods.

1. Indentation Techniques

Indentation techniques employ either a dropping weight or a constant force to deform the skin and then record the changes produced by the stress in the skin. The depth of the indentation and the time of recovery are measured to relate to viscoelastic properties. By using a dynamic system a dampening wave may be obtained that can be correlated with changes in aging. These changes can further be related to elastin and collagen alterations. In aged skin, a longer time is required for the indentation to recover, and suggests that fluid movement within the skin is impaired [8].

2. Extension Techniques

Extension techniques employ a system with a fixed bar fastened to one area of the skin and a second movable bar fixed to an adjacent area. The second bar is moved over a distance at a fixed rate to produce a stress-extension curve. While the method has shown good results, critics point out that the skin area outside the bars (or clamps) also enters into the results, making the unit area difficult to define.

3. Torsion Techniques

Torsion techniques employ a fixed ring glued to the skin and a second device fixed by friction or glue to the center area of the skin contained within the ring. This second device is rotated at a constant rate with a known force. The chief benefit of this device is that it measures shear force as well as other viscoelastic properties. The method is popular and is enjoying wider use of late, but it has inherent difficulties [11]. One of the major difficulties is the tension state of the skin at the time of the measurement. Changes in skin hydration can also have a significant effect on the results.

III. VISCOELASTIC MEASUREMENTS BY RACE-TRACK SUCTION METHOD

A. Background Information

The method described in this chapter is the method used by the writer to evaluate the efficacy of antiaging products over various application periods. Since not every reader will be familiar with viscoelastic measurements, certain basic concepts and definitions will be covered in order to provide this information.

1. Major Viscoelastic Components in the Skin

There are four major components in the skin that contribute to the viscoelastic measurements, the collagen fibers, the elastin fiber, the proteoglycans (ground substance), and water. Each of these components has a unique effect alone, but when combined with the other components a very complex response is produced.

a. Collagen. Collagen fibers are complex structures that limit the extensibility of skin. They are responsible for the large vertical portion of the curve seen in a stress-strain recording (Fig. 14.1). Collagin fibers are generally oriented along stress lines and are usually parallel when stressed to their limit.

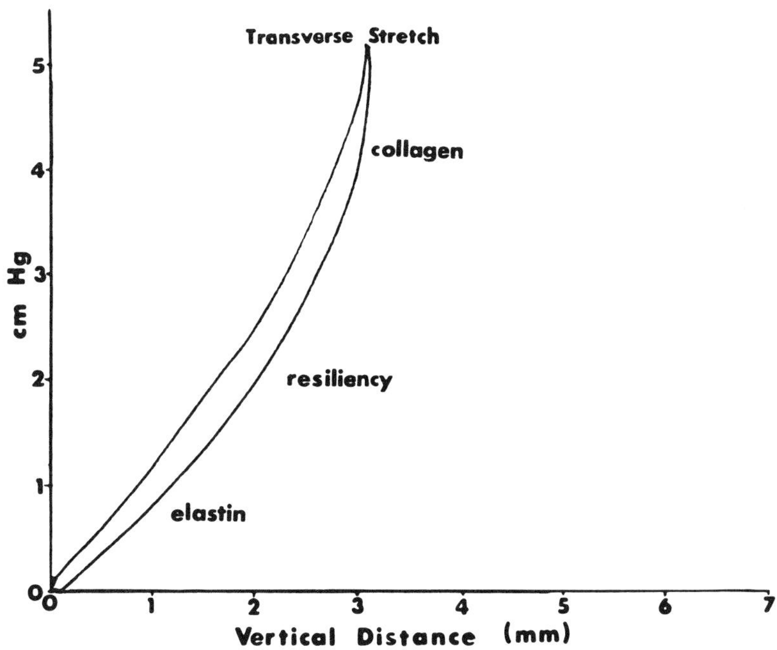

FIGURE 14.1 Viscoelastic curve showing the three components of the curve. Note the placement of the collagen component.

b. *Elastin.* Elastin fibers are entirely different chemically from the collagen fiber. Elastin provides a return spring system for the skin, allowing the collagen fibers to return to their original position after deformation. The elastin fiber component of the stress-strain record is seen as the long horizontal sloping line along the x-axis (Fig. 14.2). Elastin fibers are believed to be more randomly oriented than collagen fibers.

c. *Proteoglycans.* Proteoglycans make up the most complex component of the dermis. They are the most difficult to measure and to evaluate. On the stress-strain curve, the curvilinear portion represents the contribution made by the proteoglycans. No adequate methods are available to assay this system in the skin with in vivo techniques.

d. *Water.* Water is the single most important element in the skin that changes constantly in response to stress and strain. The nature of water bonding to the proteoglycans and the

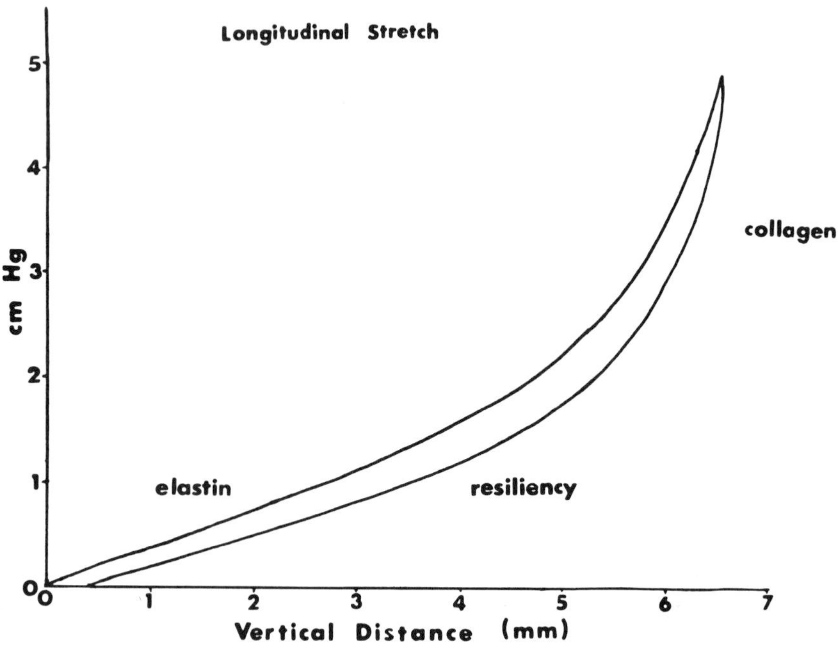

FIGURE 14.2 Viscoelastic curve showing the elastic component in the longitudinal placement of the race-track suction device. Long axis of device is placed in a parallel position to long axis of the arm.

bonding of the proteoglycans to the collagen and elastin fibers remains a fundamental area of investigation.

2. Terms Used in Viscoelastic Measurements

a. Stress. Stress is the force applied to the skin in any form. It is frequently reported as weight per unit area, or more commonly in pascals. A pascal (Pa) is 1 newton/m^2. A MPa is 1 million Pa. In case of suction devices, air pressure is used and is reported in units of torr: 1 torr = 133.32 Pa, or 1 psi (lb/$in.^2$), which equals 6894.72 Pa.

b. Strain. Strain refers to the linear distance of travel or displacement of the tissue. It is expressed in units of length, using the centimeter as the base unit. In the case of the stress-relaxation curve, the unit along the x-axis is time.

c. Modulus. Modulus is a term frequently found in the literature on viscoelastic measurements. It is essentially a ratio relating stress to strain and is a convenient method of comparing materials.

d. Hysteresis. Hysteresis is a characteristic of imperfectly elastic material, such as skin. It is seen on the stress-strain record as a second curve, and represents either a repeat stress-strain curve or a relaxation curve. The hysteresis data can be used to determine the tissue resiliency.

B. Method and Materials

1. Equipment

While a variety of equipment is available as component parts, I know of no manufacturer of a complete suction unit for viscoelastic measurements. The equipment used by the writer is readily available. The chamber, however, must be machined from plastic. The dimensions are 1.3 × 3.8 cm (see Fig. 14.3 for the general layout of the design). This race-track geometry designed by Cook [11] allows a more homogeneous stress field than the system reported by Grahame [13]. The pressure applied to the chamber is controlled by a Nullmatic pressure regulator Model 44-20 (Moore Products, Spring House, Pennsylvania). The pressure and vacuum fed into the regulator was from a standard laboratory pressure-vacuum pump (Fisher Co.).

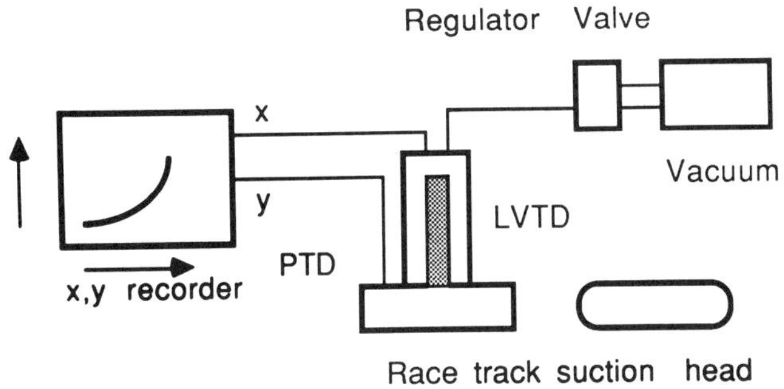

FIGURE 14.3 General layout of the suction device. PTD = the pressure transducer placed between the chamber and plugged into the y-axis. LVTD is the variable linear transducer centered in the chamber. The output the LVTD is plugged into the x-axis.

The pressure was measured with a Model XCQ-062-25G pressure transducer (KULITE Semiconductor Products, Inc., Ridgefield, New Jersey). The vertical displacement of the skin was measured with a Model 210A-0300 LVTD (Robinson-Halpern, Plymouth Meeting, Pennsylvania). Appropriate amplifiers and signal conditioners are needed, which can either be obtained from the component suppliers or adapted from existing laboratory equipment. The output of the pressure transducer is fed into the y-axis of an X-Y recorder, and the output of the linear variable transducer (LVTD) is fed into the x-axis of the same recorder. The chamber is mounted on a positioning device, which is modified from a commercial light fixture used to position desk lights. Vacuum is applied to the system at a rate of 25 cm Hg/min with a maximum of 5 cm Hg. A constant speed motor is required to operate the valve which controls the vacuum.

2. Calibration

The instrument is calibrated by moving the core of the LVTD a fixed distance with a micrometer and observing the change on the recorder. A linear relation must be found between the output of the transducer and the distance traveled. The pressure transducer is calibrated with a mercury manometer and a linear relationship obtained (Figs. 14.4 and 14.5).

3. Operation

The advantage of race-track configuration allows the user to place the unit in the long axis of the area of the body being studied, and thus to obtain specific directional information. We apply approximately 5 cm Hg of vacuum to chamber over a period of 15 sec, the vacuum is then released, and a second curve is obtained. The chamber is fastened to the skin with double-sided adhesive tape and the area tested is placed in an anatomically neutral position. It is very important to account for preexisting tension in the skin in order to obtain consistent results. While our method does not control this with a pretension device as advocated by Cook [11], we do take great care in positioning the area.

Depending on the type of product studied, the application time should extend beyond 2 weeks if any appreciable change beyond immediate hydration of the dermis is expected. We have seen no change in any parameter with this method by superhydration of the epidermis. We feel, therefore, that we are measuring only dermal parameters. Six weeks or longer are needed to detect any significant change in elastin values owing to new

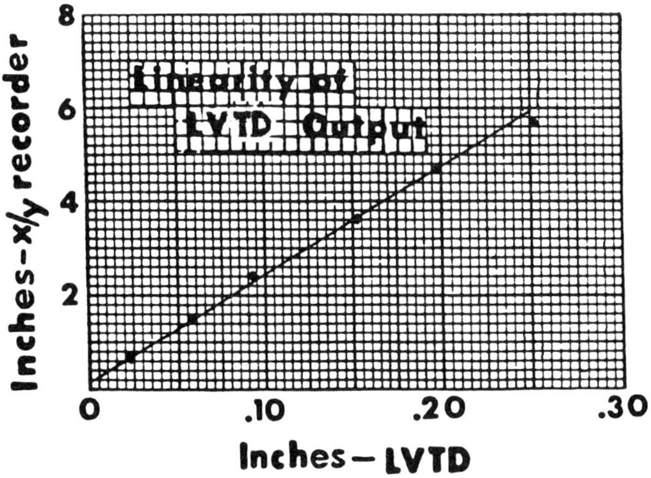

FIGURE 14.4 Linear relationship of the inches traveled on the recorder to the travel of the LVTD armature.

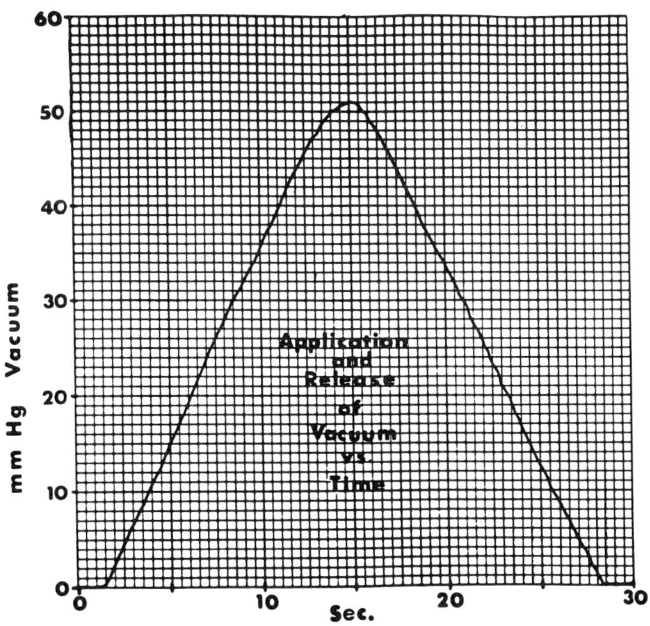

FIGURE 14.5 Relationship of the application and release of vacuum to the time frame.

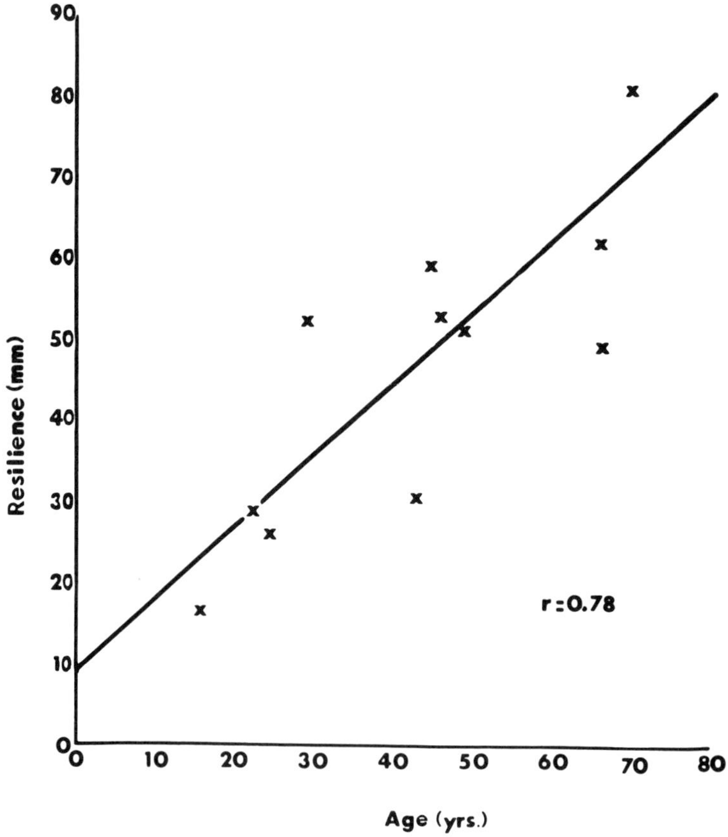

FIGURE 14.6 Correlation of resilience, defined as the greatest distance between the two curves in the hysteresis loop and subject age.

elastin formation, and probably longer for collagen changes. Histologically, changes may be subtle at 6 weeks, depending on the type of product used, but the viscoelastic parameters will be altered. We follow each subject with measurements at intervals of 2 weeks.

4. Results

The interpretation of the stress-strain curve is limited to two parameters, extensibility and resiliency in our system, though

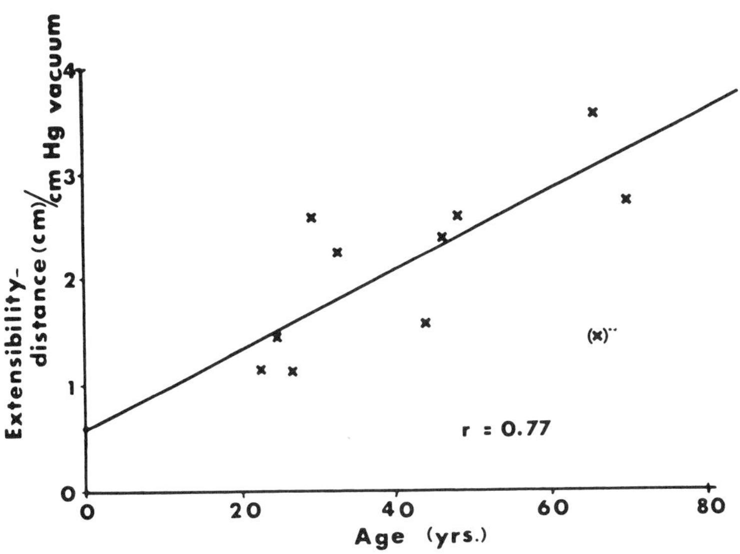

FIGURE 14.7 Correlation of the elastin component with age. The elastin component is defined as the extensibility in centimeters per cubic centimeter of mercury vacuum applied. The subject denoted by an (x) exhibited "the south Florida syndrome," i.e., had a history of extreme sun exposure over many years.

other parameters could be used. We measure the slope of the alpha line and designate this value as extensibility. The value is obtained by dividing the distance in centimeters on the graph by pressure in centimeters of mercury. The resiliency is measured as the longest distance within the hysteresis loop. This is purely an arbitrary value; one could use any point on the y-axis as long as the same point was taken for each subject (Fig. 14.6).

There are a number of points that we need to review about these types of measurements before considering their use in the evaluation of antiaging products. The spread of values obtained in a random population for any particular viscoelastic measurement is going to be fairly wide. This is not unexpected considering the number of factors that enter into aging of tissues. Figure 14.7 shows a range of values in extensibility for a small population of women from age 22 to 75 years. The correlation coefficient of 0.77 is not bad for such a small group. Next we must carefully consider the anatomical sight where the chamber is placed, as this choice can alter the results. Finally, when using female subjects who are in the child-bearing years care must be exercised to note the time of testing to the menstrual cycle in order to avoid tissue hydration artifacts.

IV. THE VISCOELASTIC SYSTEM IN EVALUATION OF PRODUCTS

The viscoelastic system has been used to evaluate antiaging products based on antioxidants, collagen stimulation, and other ingredients that promote repair and regeneration of tissue. In all cases, this is not a simple technique that will provide quick and cheap answers. Products should be tested with adequate amounts of active ingredients. In our series, we were unable to distinguish between the effects of 2-3% of an active versus 3-5% of the same active ingredients, since in many cases an all-or-none effect is seen. The "noise level" of this method in our hands appears to be between 5 and 10% change over the control or untreated value. Any change below 10% is considered non-significant. Finally, we feel an adequate trial of at least 3 months of product usage should be considered before you abandon the product as ineffective. A reasonable guide to follow is the subject's appraisal of the product's effect and your own subjective appraisal of whether the subject appears to have improved or not improved. There is usually good correlation between the clinical evaluation and the objective value obtained by the viscoelastic measurements.

V. MEASUREMENT OF SKIN HYDROPEROXIDES

The skin surface lipids are very prone to attack by oxygen both from the atmosphere and from biological oxidation processes. It is possible to assay the level of peroxidation of these lipids in a noninvasive manner. This technique is described in detail in this section [14].

A. Extraction Procedure

Skin lipids may be safely extracted from the skin using acetone as the solvent. A glass cylinder measuring 5 cm in diameter is placed on the skin and held snugly. Two aliquots of pure acetone are used to extract the lipids in 5-ml portions. The first

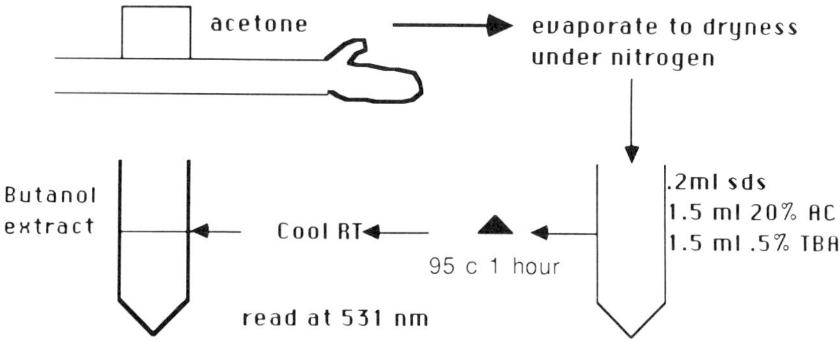

FIGURE 14.8 Flow diagram of the skin surface lipid assay.

5 ml is added to the cylinder and the skin is gently rubbed with a Teflon rod for 1 min. The acetone is removed with a Pasteur pipette and placed in a test tube. A second 5-ml portion of acetone is added to the cylinder and the skin is again gently rubbed with the Teflon rod for 1 min. This second portion of acetone is added to the test tube along with the first and the sample is dried under a nitrogen stream (Fig. 14.8).

B. Thiobarbituric Acid Reaction

The lipid residue in the test tube is emulsified with the following mixture: 0.2 ml of 8% sodium dodecyl sulfate, 1.5 ml of 20% acetic acid, and 1.5 ml of 0.5% thiobarbituric acid solution. The final volume is made up to 4 ml with water. The tube is then heated to 90°C for 1 hr and then cooled to room temperature. After cooling the tube is centrifuged and supernatant is extracted with 4 ml of n-butanol. The absorption is read at 531 nm and the final concentration of peroxides is determined as malonyldialdehyde (MDA) precursors. Calculations are as follows:

$$\text{nanomoles of MDA} = \frac{\text{optical density} \times \text{volume}}{0.15}$$

where V is the final volume in milliters and 0.15 is the extinction coefficient of MDA at 531 nm (using standard EC values). MDA standards may be prepared from 1,3,3,tetraethoxypropane.

TABLE 14.1 Acetone Extractable Lipid Peroxide from Forearm Skin

Subject age	MDA in nm/100 mg lipid
42	0.41
49	0.57
50	0.51
55	0.68
62	0.94
71	1.18
79	1.45

C. Results

There is a linear relationship between the quantity of peroxides recovered and the subject's age. This study may be used to evaluate products when used over time. Since the production of peroxides are markedly influenced by ultraviolet light, care must be exercised in long-term studies regarding the exposure of subjects to this agent. Table 14.1 shows the results of a small number of subjects randomly selected for this assay without prior preparation. In this assay, the measurement was made per 100 mg of extracted lipid. Total lipids are assayed by the vanillin method [15].

REFERENCES

1. Rosen, R., Dynamical Aspects of Senescence. Chapter 2 in Biological Mechanisms in Aging, National Institute on Aging, Publication No. 81-2194, 1981.
2. Hayflic, L., and Moorehead, P. S., The serial cultivation of human diploid cell strains, Exper. Cell. Res., 25:585-621, 1961.
3. Good, P. I., Aging in mammalian cells population—Review, Mech. Ageing Dev., 4:339-348, 1975.
4. Smith, J. R., and Lumpkin, C. K., Loss of gene repression activity: A theory of cellular senescence, Mech. Ageing Dev., 13:387-392, 1980.
5. Cutler, R. G., Evolution of longevity in primates, J. Human Evol., 5:169, 1976.

6. Bjorksten, J., The crosslinkage theory of aging, J. Am. Geriatr. Soc., 16:408-427, 1968.

7. Daly, C. H., and Odland, G. F., Age related changes in the mechanical properties of human skin, J. Invest. Dermatol., 73:84-87, 1979.

8. Kydd, W. L., Daly, C. H., and Nansen, P. D., Variation in the response to mechanical stress of human soft tissue as related to age, J. Prosthet. Dent., 32:493-500, 1974.

9. Evans, J. H., and Sieawnop, W., Controlled quasi-static testing of human skin in vivo, Digest of the 7th International Conference on Medical Biological Engineering, Stockholm, 1967.

10. Finlay, B., Dynamic testing of human skin in vivo, J. Biomech., 3:557, 1970.

11. Cook, T., Alexander, H., and Cohen, M., Experimental method for determining the 2-dimensional mechanical properties of living human skin, Med. Biol. Eng. Comput., 15:381-390, 1977.

12. Bazin, R., Potart, J. L., Pascual, I. Her. C., Schmidt, C., and Soudant, E., Biomechanical skin results obtained by a suction method. Preprint XIVth I.F.S.C.C. Congress Barcelona, 1:211-224, 1986.

13. Grahame, R., A method for measuring human skin elasticity in vivo with observations on the effects of age, sex, and pregnancy, Clin. Sci., 39:223, 1970.

14. Ohkido, M., Yoshino, K., and Matsuo, I., Lipid peroxide of human skin, Curr. Prob. Dermatol., 10:269-278, 1980.

15. Toro, G. T., and Ackerman, P. G., Practical Clinical Chemistry, Little Brown, Boxton, 1975, pp. 352-355.

15

Salon Testing

Linda A. Lazar

Consultant, Bloomfield, New Jersey

I. INTRODUCTION

Salon testing of hair preparations is a way to gain valuable information on specific product performance. New information is obtained each time a product is tested, whether it be a new product, or a product that has already been marketed and has had its formulation altered.

One reason for salon testing is claims substantiation: to verify that the product(s) will not sting the eyes, or burn the scalp; that the product does help mend split ends, detangles hair, increases manageability, decreases fly-away, or increases sheen.

Today, the average consumer has limited free time to frequent beauty salons and has come to depend on prepackaged hair-treatment kits, which are readily available at their local stores. When consumers purchase a shampoo, conditioner, hair dye, permanent wave, or any hair preparations that are available, they want to be confident that these products will perform as promised.

Salon testing of these products can help assure the manufacturer that the products they sell will perform safely and effectively. The more data about these products that is obtained, the more information can be passed on to the consumer. The consumers, in turn, can select products that are best suited for their needs.

II. FACILITY

The testing facility should be a large enough area to house all the equipment you need to work in efficiently. A salon sink and reclining chair should fit in the area comfortably with room to walk around on both sides. The flooring should be of a nonporous material to make it easy to clean up any spills and to avoid staining. Adequate lighting is very important. Temperatures should be 70-75°C, to give a comfortably warm climate that prevents subjects with wet heads of hair from becoming chilled. The beauticians' work stations should also be constructed of a nonporous material and have sufficient storage to hold all of the supplies you will need for each subject. The area should be well ventilated in case you are working with test materials that give off an odor. The surroundings should be made as comfortable as possible for the subject. A few current magazines should be available, especially if the procedure will take any length of time.

III. EQUIPMENT

Your equipment must include the following:

1. Salon sink and reclining chair
2. Large wall mirror
3. Generous-sized work station with a large counter area
4. Hair dryers, blow dryers, curling irons
5. Assorted sized combs, brushes, clamps or clips to section hair, aluminum foil
6. Towels, shampoo, and conditioners
7. Capes to protect the subjects' clothing
8. Rubber gloves
9. Scissors, perm rods, end papers, and rayon-reinforced coil with a handy dispenser top
10. Vaseline (to use around hair line)
11. Timers

A towel bin or hamper to store soiled towels and a disinfectant, sterilizing solution to clean counters and other equipment are also a necessity.

IV. SUBJECT SELECTION

Subjects are selected from a group of prescreened volunteers in the community. They come in all shapes and sizes and from all walks of life. Some are professional people who enjoy participating in studies, and some are housewives, cab drivers, or college students. Their names are selected from our computer, to which information was given from a questionnaire filled out on their first visit to the facility. For each study there is a protocol, which specifies what requirements must be met to enable a subject to qualify for the study. They may need to have long hair, short hair, straight hair, curly hair, dyed, gray, or permed hair. Study requirements may specify males or females for the study, or both. The selected panelists are called to see if they would like to participate. The procedure to be used is explained, and if they are interested in participating, they are instructed when they are to report to the test facility.

When the panelists arrive at the test facility they read and sign a consent form. A medical history form is filled out, and their scalp is examined by the beautician and, if needed, a board-certified dermatologist and/or other medical personnel. It is then determined whether that particular subject is qualified for the study.

V. CRITERIA

Before a subject can be empanelled for a test, there are certain criteria they must meet.

A. Inclusion

To qualify the subject must:

1. Be male or female (depending on test) and in good health
2. Fit the age criteria indicated in the protocol
3. Have a moderate amount of hair (shampoo or perm studies)
4. Have skin that is lightly pigmented (unless specified otherwise)

B. Exclusion

The subject would not qualify if the following conditions exist:

1. If they have participated in any similar test in the last 2 weeks

2. If they have any skin disease or scalp irritations
3. If they are taking any medications
4. If they are allergic to any hair-care/shampoo products

VI. FORMS

A. Consent Form

Each subject must read and sign a Consent Form. This form explains the purpose of the study, the procedure that will be used, and if any reaction(s) are anticipated. They are advised that they may withdraw from the study at any time and that if they have an adverse reaction, the investigator may discontinue them from the study. Subjects are also informed on any benefits they may derive from the study. They are informed of how many subjects will be participating in the study, and that their records will be kept confidential. They are also informed that medical treatment will be given at no cost for any product adverse reactions that occurred. A phone number is given in case the subject has any problems. The form is signed by the subject plus a witness and is dated.

B. Medical History Form

A medical history form is completed for each subject. Data taken are their age, birth date, sex, and race. They are asked if they have any skin disorders, and if the answer is yes, what they are. Also, if they have any allergies to shampoo/hair-care products.

Subjects are asked if they have any other significant medical problems (lung disease, cancer, high blood pressure, etc.). If they are taking any medications, and if the answer is yes, which medication and why they are taking it.

Only after all the forms are read and signed, the medical history taken, and the scalp examination completed are the subjects empanelled in the study.

VII. PREPARATIONS

A. Half-Head Tests

1. Shampoo

When you want to test two different products at the same time you do a half-head test. With a shampoo study, products are applied to each half of the head. One can be the test product and the other is usually a control product. Alternate sides of the head are randomly used on each subject. Predesignated bottles of shampoo (by subject number) which have been randomized as to the half-head location by the company are used on the subjects. The number of subjects used is varied, but usually consists of approximately 30.

With a one-product study, the procedure is shampoo, rinse, and repeat, with a designated amount of shampoo (depending on the length of the hair and its thickness). The hair is rinsed thoroughly between shampoos, and especially at the final rinse. Evaluations are made for combing resistance while the hair is wet. Hair can then be dried, and further evaluations are made. For example, is the hair controllable? Does it have body and shine?

Sometimes an evaluation for the lathering abilities of a shampoo is required (the average consumer looks for a shampoo with a good lathering ability), especially for oily hair. They also look for a shampoo that rinses out well, leaving the hair feeling clean.

If an evaluation of a dandruff shampoo is required, the subjects are shampooed with an inactive (placebo) shampoo for a period of about 1 week. An assessment is made of:

1. Area affected with dandruff.
2. Severity of flaking.
3. Is dandruff loose, or does it adhere to the scalp?
4. Corneocyte count.

These assessments are made by a dermatologist or the other trained medical personnel. The subject may be asked about the use patterns of the products they previously used. The subjects are told that the test product is the only product that will be used for the duration of the study.

Only healthy adult males or females who have a moderate to severe case of clinically diagnosed dandruff will be used for the study. Subjects may return to the salon twice a week for supervised shampoos (for approximately 8 weeks), and their scalps are reexamined every 2 weeks by the determatologist. After each shampoo, the subjects are asked if their scalps tingle, burn, or sting. Their answers are entered on a subject test log. The condition of each subject's hair is evaluated at the beginning of the study, again at designated intervals, and at the conclusion of the study by an experienced beautician.

Because some antidandruff formulations are often considered a drug, there may be the need for an Institutional Review Board (IRB) approval. Five members of the board meet and discuss the procedures being used and the formulations. If IRB approval is denied, the study cannot take place. This procedure is done to protect the rights of the subjects.

At the conclusion of the study, the subjects can be asked to answer a questionnaire. They are usually asked if they thought their hair was controllable after they were shampooed. Also, if they thought the shampoo controlled their dandruff. In general, did they care for the shampoo, and would they purchase the product. They will also be able to add any other comments they might have about the shampoo product.

2. Hair Coloring

Hair coloring is done by seating the subject in a reclining chair, attaching a towel and cape around their neck (to protect their clothing), then applying the dye to dry hair. (Vaseline is put around the hair line, which is especially important when using darker shades to prevent staining of the skin.) If only one product is being used, the hair dye is applied to the entire head. If it is to be a half-head test to compare two products, one product is applied to the right side, and the second product is applied to the left side of the head.

When there is a one-product test, the subject is best seated in a reclining chair with his/her head resting on the salon sink so that he/she is comfortable. The operator must have easy accessibility to rinse water.

There are two different types of hair dyes:

1. Semipermanent dyes, which are those that last only through five to six washings (considerable fading). These dyes are used when one wants to highlight or add body to the hair (a more subtle change).
2. Permanent dyes, which last about a 6-week period (very little fading), after which a new application is required because of new hair growth. This type of hair coloring is also used to change the hair color more than two shades lighter or darker (a more drastic change).

Because human hair is more porous on the tips than the root area, and although hair dyes can penetrate the root and ends, the ends of the hair wash out much faster. This can be so much faster that at times a lot of the hair color is washed out during the rinsing stage of the coloring process. Dyes are left on the hair from 10 to 30 min.

Evaluations are made at this time as to how much body, shine, and manageability the hair has. Also when the hair is dried, does it retain most of the color?

B. Permanent Waving

Permanent wave treatments are performed on subjects with virgin hair (hair that has no hair dyes or permanent waves). Perms are performed in the standard manner. Rolling of the hair in a permanent wave is one of the most important factors in getting good results from a wave. Hair is sectioned and secured with clips or hair clamps. Rolling is begun at the nape of the neck. Uniformed sections of the hair are then wrapped in end papers, and wound around a permanent rod. (Size of rods determines the size of the curl; the smaller the rod, the tighter the curl.) A rayon-reinforced coil is circled around the hair line and nape of neck to catch any drippings from the perm solution. Hair thickness is a determining factor in the timing process of the permanent wave. Coarse hair is resistant and requires a longer timing period. It also usually requires a higher strength waving solution. With fine hair you will need less of a timing period and a weaker solution.

Evaluations are made as to the degree of curl, curl retention abilities, and the cosmetic appeal of the hair, in addition to any evident hair damage.

C. Hair Damage

There are many ways one can cause damage to the hair. The most frequently used styling aids are blow-dryers, curling irons, and electric rollers. All these aids, in addition to hair dyes, permanent waving, etc., can cause significant damage to the hair, not to mention the natural factors in the environment (such as sun exposure).

Improper combing and brushing of the hair, especially when the hair is wet, is extremely damaging because these cause the hair to catch in the bristles; the hair is then stretched, becomes tangled, and results in split ends. Constant blow drying and the use of curling irons also result in causing dry, brittle, split ends, which can produce severe breakage, leaving the hair looking very unattractive.

Most of the preparations manufactured today are formulated to help reduce, if not eliminate, dry, split ends. Creme rinses and conditioners are products formulated to help this problem. Evaluations made after these products have been tested determine if the split ends have improved, making them less visible. This is especially owing to the fact that there is always more damage to the ends than the root area.

D. Nail Enamel

What consumers look for in nail enamels:

1. Easy to apply
2. Are fast drying
3. Have a good shine
4. Chip resistant
5. That they are nontoxic

The basic manicure consists of a base coat, two coats of the enamel, and last a top coat (sealer). Nail enamels are available in creme or pearl (frosted) shades.

Nails are polished by a professional manicurist with both a test and a control enamel using alternate fingers. Polish is usually worn for 1 week and examined daily. The condition of the enamel is recorded on data sheets. Participants are asked how often and how long a time their hands are immersed in water.

The most reliable testing of nail enamels is applying it to the nails. If the subject is to apply the enamel herself, then the evaluation would include if she thought the enamel applied evenly, without streaking. This evaluation is done at the time of application of the nail enamel.

VIII. PARAMETERS

Listed below are various parameters that are used on a subjective graduated scale for salon evaluation of hair.

Combing Resistance

0 No resistance
1 Minimal resistance without snag (snarl)
2 Moderate resistance with occasional snag
3 Many snags (snarls) with frequent comb stoppage

Static

0 Lays in place without lift
1 Minimal static, but controllable

2 Moderately charged, somewhat uncontrollable
3 Highly static, fly-away, difficult to control

Split Ends

0 None
1 Minimal, trace amounts, difficult to observe
2 Moderate, some breakage, easy to observe
3 Severe breakage, brittle, damaged, observed from distance

Volume/Body

0 Limp, droopy, shapeless, no depth
1 Some bounce, double volume (thickness)
2 Moderate bounce (fluffy), triple volume (bulk)
3 Maximum bounce (fullness), quadruple or more

Manageability

0 Nonmanageable
1 Somewhat manageable with considerable effort
2 Somewhat manageable with little effort
3 Completely manageable (easy to style)

Sheen/Shine

0 Very dull, lacks luster, dry, lifeless
1 Minimal shine, with intermittent dull areas
2 Shine with some highlights
3 High luster, glossy, silky reflect

APPENDIX

INCLUSION / EXCLUSION CRITERIA

Subject Number ________ Subject Initials ________

<u>Inclusion Criteria</u>

	-Yes-	-No-
a. Is male or female in generally good health.	____	____
b. Is between the ages of 18 and 65.	____	____
c. Has not participated in a similar clinical investigation during the past 4 weeks.	____	____
d. Is not bald (has a reasonable amount of hair).	____	____
e. Has lightly pigmented skin.	____	____
f. Has not used a medicated (anti dandruff) shampoo in the past 2 weeks.	____	____

A NO response to any of the above criteria will preclude this subject from participating in this test.

<u>Exclusion Criteria</u>

	-Yes-	-No-
a. A known history or present condition of allergic response to shampoos or conditioner products.	____	____
b. Skin disease (e.g., psoriasis, atopic dermatitis or other cutaneous manifestations on the scalp), which would interfere with the test readings.	____	____
c. Medication (e.g., steroids or antihistamines), which would compromise the study.	____	____

The above criteria <u>A THRU C MUST BE ANSWERED NO</u> in order for the subject to participate in this test.

	-Yes-	-No-
1. Has the subject completed a medical questionnaire and is he/she in good health?	____	____
2. Has the subject been informed of the nature of the study and signed an informed consent document?	____	____
3. Does the subject satisfactorily meet all selection criteria for enrollment?	____	____

______________________________ ________

Investigator Signature Date

APPENDIX 15.1

Subject No.:__________
Subj. Initials:________

INFORMED CONSENT

Evaluation Of A Medicated Shampoo
Containing Conditioners

The clinical evaluation for which you have volunteered will recruit 6 subjects and is designed to evaluate the efficacy of a medicated shampoo containing conditioners.

Prior to admission to the study, you wil be asked several questions concerning your medical history. You will also receive a baseline evaluation of your scalp and hair.

You will then have your hair washed with the test shampoos (supplied by the Clinic). Your hair will be parted down the middle. The Beautician will use one of the coded shampoo formulas for one side of you hair and the other formula for the other side. She will then do a comb out of your hair, blow it dry, and do another comb out. During these comb outs your hair will be evaluated on each side for several management characteristics.

Severe reactions are not anticipated in this minimal risk investigation, although some subjects may experience some temporary irritation such as redness, itching, burning or stinging. If experienced, these reactions will be recorded. Any treatment required for a side effects due to study material will be provided at no cost to you, but you are not entitled to monetary compensation for such effects. If you have any questions or experience any difficulty during the study, please contact the Investigator of the study.

Benefits from the research will include new information on the manageability of the hair after using the shampoo. You will be paid a designated amount of money for your participation in this investigation. If you do not complete the investigation, no payment will be made. However, if the Investigator(s) decide that you should no longer participate (because of irritation or some other reason), you will be compensated proportionally for the time that you were involved.

Participation is voluntary and you may withdraw at any time for any reason. The Investigator may stop your participation without your prior consent for any reason including an adverse reaction to the test product, a change in the testing procedure, or your failure to comply with the instructions.

APPENDIX 15.2

Confidentiality will be maintained by the testing facility. However, the Food and Drug Administration may conduct an inspection,which may include a review of records identifying you as a subject.

SUBJECT'S DECLARATION

I represent that I am at least 18 and not over 65 years of age; and that I have not participated in an investigation in which shampoos or conditioners have been tested in this or any other manner within the past four weeks.

I have read the description above concerning this investigation and all my questions have been answered.

I also represent that, to the best of my knowledge, I do not have an allergy to shampoos or any other hair care products.

I may receive a copy of this consent form upon request.

Subject Signature:______________________Date:____________

Witness Signature:______________________Date:____________

Subject's Phone Number____________________

APPENDIX 15.2 (continued)

MEDICAL HISTORY FORM

Date: ________________
Subject No.: __________

Initials:______ Age:_____ Birthdate:___/___/___ Sex:_____
Race: Caucasian ( ) Black ( ) Hispanic ( )
Other ( )________________________________

Dermatological Disorders: Eczema___ Psoriasis___ Dry Skin___
Irritations or rash on scalp?___
Other____ ☐ None

Allergies: Hayfever___Asthma___Urticaria___Contact Dermatitis___
Cosmetics___ Medications (type)______________________
Deodorants/Antiperspirants; if so, which one(s)?__________
Other____ ☐ None

Other Significant Medical Problems: Lung Disease___ Diabetes___
Cancer___ High Blood Pressure___ Heart Disease___ Ulcers___
Liver Disease___ Kidney Trouble___ Thyroid Disease___
Rheumatic Disease___ Gall Bladder___
Other (explain)______________________________________
____ ☐ None

Are you taking any medication? ___ No ___ Yes, detail what medication and why: ____________________________________
____
____

________________________________ ______________
Investigator's Signature Date

APPENDIX 15.3

Investigator's Case Report Form

Subject No. ______________ Subject's Initials__________

Erythema Scale

Scalp Findings:

0 = No erythema
1 = Mild erythema, slightly pink, not diffuse across scalp, mottled, patchy
2 = Moderate erythema, definitely red, diffuse across scalp, with occasional excoriation
3 = Severe erythema, diffuse over area of involvement, excoriation and/or scabbing

	Date	Score	Comments
Initial Visit			
1st Treatment			
Week 1			
Week 2			
Week 4			

General Comments:____

____

____

____

_________________________________ ____________________
Investigator Date

APPENDIX 15.4

BEAUTICIAN'S REPORT FORM

Subject Number________ Subject Initials________ Sex F M

Age______ Grey_____ Bleached_____ Colored_____ Permed_____

PARAMETERS

Manageability
0- nonmanageable
1- somewhat manageable with considerable effort
2- somewhat manageable with little effort
3- completely manageable (easy to style)

Combing Resistance
0- no resistance
1- minimal resistance without snag
2- moderate resistance with occasional snag
3- many snags with frequent comb stoppage

Sheen/Shine
0- very dull, lacks luster, dry, lifeless
1- minimal shine, with intermittent dull areas
2- shine, with some highlights
3- high luster, glossy, silky reflection

Static
0- lays in place without lift
1- minimal static, but controllable
2- moderately charged, somewhat uncontrollable
3- highly static, fly away, difficult to control

	Manageability	Sheen/Shine	Combing Resistance	Static
Initial Visit				
1st Treatment				
Week 1				
Week 2				
Week 4				

APPENDIX 15.5

Comments: ______________________________

____

____

____

Signatures

______________________________ Beautician ______________ Date

______________________________ Investigator ______________ Date

APPENDIX 15.5 (continued)

Subject# ________

Initials ________

Week: 1 2 3 4

SUBJECT DIARY

Instructions for recording comments on the following log:

1. Use this form to record any comments.
2. Record any positive or negative factors on the day they occur.
3. Please wait until just prior to retiring for the evening before you complete the form.

COMMENTS

DAY 1 ____________________

____

DAY 2 ____________________

____

DAY 3 ____________________

____

DAY 4 ____________________

____

DAY 5 ____________________

____

DAY 6 ____________________

____

DAY 7 ____________________

____

Completed by: ____________________
(Subject's written initials)

For any concerns, please contact:
The Investigator of the study.

APPENDIX 15.6

Index